'Everything should be made as simple as possible, but not simpler.'

— attributed to Albert Einstein.

Other titles from the Simplicity Research Institute

Integrated Mathematics for Explorers
by Adeline Ng and Rajesh R. Parwani

Real World Mathematics
by Wei Khim Ng and Rajesh R. Parwani

Available from the online SRI Bookstore
www.store.simplicitysg.net
and other outlets

Simplicity in Complexity

An Introduction to Complex Systems

Rajesh R. Parwani

Simplicity Research Institute, Singapore
www.simplicitysg.net

Simplicity in Complexity
Published by the Simplicity Research Institute, Singapore
www.simplicitysg.net

For bulk orders, special discounts, or to obtain customised versions of this book, please contact the Simplicity Research Institute at **enquiry@simplicitysg.net**.

A CIP record for this book is available from the National Library Board, Singapore.

Print Edition SRI-2015-1A
ISBN: $978 - 981 - 09 - 3932 - 8$ (pbook)
ISBN: $978 - 981 - 09 - 3933 - 5$ (ebook)

Contents

Preface

A working hypothesis of scientists is that there exist relatively simple causes underlying all phenomena, no matter how complex.

Roughly speaking, one says that a system is complex if it consists of many interacting components. The complexity that is usually observed arises from the size of the system (the number of components) rather than the interaction rules.

The title of this book summarises the two ways that simplicity[1] can manifest itself: Simplicity of the underlying rules of a complex system and simplicity of emergent laws and structures.

We will see how relatively simple rules, principles and methodologies can be applied to study systems ranging from the domain of physics to other fields such as chemistry, biology, ecology and sociology.

In this book, we will not focus on a detailed study of individual complex systems, but on uncovering common themes that can be used in our understanding of different systems. This book may, therefore, be used for an introductory course on Complex Systems.

The exercises at the end of each chapter are meant to encourage active learning through reading, thinking, experimentation, and discussion with other emergent beings.

We welcome feedback and questions from users of this book; please email us at **enquiry@simplicitysg.net**. Updates will be posted on the book's webpage at

www.simplicitysg.net/books

Some parts of the 'Model Building' and 'Dynamical Systems'

---

[1]'Simplicity' does not necessarily mean 'easy to understand'. However, in this book we will attempt to keep things easy.

chapters had previously appeared in another book by one of the current authors: 'Real World Mathematics' by W.K. Ng and R. Parwani.

0.1   Conventions

Key concepts are highlighted in italics while footnotes provide clarification or commentary. Single 'quotes' focus on particular words.

Where possible, we have attempted to locate the original sources for the ideas summarised in each chapter, and placed those references at the end of the chapter together with accessible secondary references[2].

0.2   Acknowledgements

I am grateful to the hundreds of students who participated in the 'Simplicity' and 'Complexity' modules between 2001 and 2012, helping to evolve the material that has been partly condensed into this book.

I am also grateful to Claudio Coriano, Poonam Dadlani, Lingzhi Li, Sergio Mendoza, Wei Khim Ng, Adhiraj Saxena, Parizad Setna, Zhuo Bin Siu, Gwendolyn Regina Tan, Yingrou Tan, Linda Toh, and Qinghai Wang for providing helpful feedback on parts of the draft manuscript.

Rajesh R. Parwani

Jan 2015, Singapore.

---

[2]Readers who notice any omissions are welcome to provide us with feedback.

Chapter 1

Introduction

1.1 What is a Complex System?

We use the word *system* to refer to the limited part of the universe that we wish to study. Hence the universe is conveniently divided into two parts, the system and an exterior *environment*.

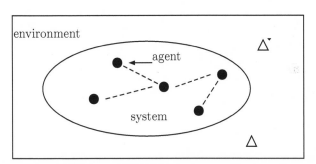

Figure 1.1: System and Environment.

Examples of systems are: An ant colony, a gas in a box, the stockmarket, a gene network, and traffic flow along a road network.

Most systems are *open*, allowing for interaction (an exchange of energy, matter or information) between the system and the environment; some systems may be approximated as *closed* (or *isolated*).

Each individual component of a system is referred to as an *agent*. An example of an agent might be an ant, an individual stock buyer, or a gene.

A *complex system* typically has a large number of interacting agents, and is often of interest because it displays some patterns that are not an obvious consequence of the underlying rules. The goal would be to understand, through a detailed analysis, how those patterns follow from the rules[1].

1.2   Overview of Complexity

Complexity refers to the study of complex systems[2].

What are some of the common characteristics[3] of complex dynamical systems?

One often quoted concept is that of *emergence*, which refers to the appearance of novel laws, patterns or order through collective effects: The emergent phenomenon is not an intrinsic property of the components, but rather something that is a feature of the system as a whole. Simple examples are those of 'temperature' (of a gas in a box), and the 'Central Limit Theorem' of statistics: At the single component level those concepts make no sense, being collective features of a large system.

Sometimes one sees the phrase 'the whole is more than the sum of its parts' as a definition of emergence. This reflects the typical non-linearity of complex systems, whereby the output is not proportional to the input, and small changes can give rise to large effects.

It is important to realise that the universe consists of many *hierarchical levels of complexity* linked to each other. Each level has its own emergent patterns and laws: As one goes up from quarks, atoms, molecules, cells, organs, organisms, ecosystems, planets, solar-systems, to galaxies, different effective laws emerge.

However, those laws would not be useful if there was not some degree of *universality*, that is, one hopes that at each level of complexity the same laws apply to varied systems rather than each following its own tune. For example, quantum mechanics is needed at the atomic level, but larger systems are well described according to Newtonian laws, while engineers often use empirical rules, as do social scientists.

---

[1]In contrast to 'complex system', the phrase 'complicated system' suggests something that would be difficult to understand. One way of contrasting those two notions is suggested in an exercise in Chap.(10).

[2]The phrase 'Complexity Science' is sometimes used instead of 'Complexity'.

[3]The key concepts will be elaborated in later chapters.

It is the apparent universality of the laws of physics, for example, that makes the world comprehensible and gives us faith in its ultimate simplicity.

––––––

Living systems are the most complex examples one can think of, and it is remarkable how such systems tend in their development towards greater order. One can say that living organisms have a tendency to *self-organise.*

Ant colonies are classic examples of self-organisation. Without a leader orchestrating everything (the queen mainly lays eggs), each ant seemingly does its own thing, apparently following a few simple rules that determine its interaction with its environment or its ant mates. Yet, an incredibly organised and sophisticated society emerges.

Ant colonies are also fascinating examples of *complex adaptive systems*, able to adjust to changing circumstances, using both *feedback* mechanisms and parallel analyses of options. Ant colonies serve as an inspiration for not only social scientists, but also computer scientists who wish to design algorithms tackling difficult problems.

However, not all systems in nature appear organised or manifest an obvious pattern. Indeed, many seem to be driven by random events. However, some of that randomness might only be on the surface. *Chaos* refers to the property of some deterministic[4] *non-linear* systems that are extremely sensitive to initial conditions and display long-term aperiodic behaviour that seems unpredictable.

––––––

Sometimes one encounters debate between approaches that are labelled *reductionist* or *holistic*. Generally, reductionists are more interested in the basic components that make up the whole, and the corresponding inter-component rules: To them, the behaviour of the whole is a consequence of the rules among the components. Particle physicists[5] and molecular biologists usually fall into the 'reductionist' category.

While it is undoubtedly true that knowledge of the components of a system, and the basic interactions among those, is essential for

––––––––––––––––––––

[4]Precisely defined, with no randomness.

[5]Particle physicists study the fundamental constituents of matter.

us to progress, it is also a fact that such knowledge by itself is often insufficient to predict all the novelty that can arise in a large system. For example, it took much effort after the discovery of superconductivity[6] to understand how it arises as a collective effect — one knew what to look for in the equations. Similarly, knowing the whole genetic code is not going to predict for us every feature of an organism or a society[7].

The problem of precisely deducing the whole of a large system from its parts is at least two-fold. Firstly, it is a computational problem. Systems with a large number of variables are too complicated for exact solutions, and probabilistic averaging methods have limited applicability.

The growth of computer power at low cost has allowed large systems to be simulated or solved numerically. However, this highlights the second problem: Often one does not have full knowledge of the fundamental dynamics, or the initial conditions, or the problem is still too complicated to be handled directly by computers at present.

Often some guesswork or intuition is required to reduce the actual problem to a simpler *model* which can then be tested on a computer. *Computer simulations* of simplified models let one test assumptions quickly, and if the results appear similar to the real world, then one can tentatively accept the validity of the model and its assumptions. Qualitative similarities, of course, do not constitute a proof, because other models with different assumptions might give similar results, but at least the insight gained helps one to make further guesses and tests in a particular direction rather than being lost in a mess of detail.

A valuable lesson computer simulations have taught us is that a large system with very simple local rules can give rise to collective behaviour of great complexity and variety, showing, on the one hand, that complex phenomena need not require complicated rules, but at the same time reminding us how difficult it is (without computers) to deduce the emergent behaviour from the components and their interactions.

Thus studying the whole is as interesting as studying its parts,

---

[6]A superconductor is a material that has zero electrical resistance below some low critical temperature. Current interest is in 'high-temperature superconductors', for which a theoretical understanding is still lacking at the time of writing.

[7]In the foreseeable future.

as novel structures and emergent laws arise at each hierarchical level of complexity. The specialist physicist studying superconductivity is not going to be replaced by the particle physicist, and neither is the ecologist going to be become obsolete because of the molecular geneticist[8].

1.3 Summary

This book aims to emphasize how an approach based on quantitative models and the scientific method can give useful trans-disciplinary insights into many interesting systems ranging from the disciplines of physics to the social sciences.

Bear in mind that there are differing personal perspectives and emphases in the inter-disciplinary study of complex systems, some areas of which are very much at the frontier of research.

1.4 Exercises

1. If the human heart is the system of interest for a study, what might the agents be? What would be the environment?

2. If the human heart was an agent in a system, what might the system be? What would be the environment?

3. Are the systems in the above questions open or closed?

4. An engineer is interested in studying traffic flow along a highway. What would be the system, the agents and the environment?

5. A cosmologist is a scientist who studies the whole universe. What would be her system, agents and environment?

---

[8]A molecular geneticist studies genes at the molecular level.

"For the things we have to learn before we can do, we learn by doing". — Aristotle[9]

---

[9]Source: en.wikiquote.org; as cited in the Oxford Dictionary of Scientific Quotations (2005), 21:9.

Chapter 2

Model Building

Our understanding of the world is facilitated by building models. In this chapter we briefly discuss the model building philosophy and the importance of the scientific method.

2.1 The Scientific Method

A scientist seeks to find the most economical and accurate description of phenomena using the *scientific method*: Testing hypotheses and theoretical models against data from experiment or observations of actual phenomena.

The word *model* usually refers to a tentative attempt to explain a phenomenon or data. Once a model has been well-tested and developed to sufficient generality, it is often called a *theory*[1].

The scientific method is not a linear process but involves many interlinked and iterative steps in the search for understanding. Typically, one first has some phenomenon in need of an explanation. The phenomenon might exhibit some regularities that may be summarised by some empirical relations. Experiments (or observations) might be conducted to check the robustness of the phenomenon under controlled conditions. A *hypothesis* may then be formulated: This is just a guess as to the cause of the phenomenon. Predictions can be made based on the hypothesis and further experiments conducted to test them.

---

[1]The terminology is, unfortunately, not standard.

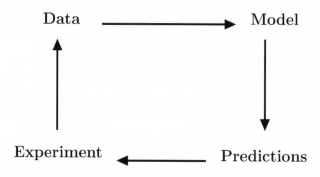

Figure 2.1: Ingredients in the scientific method.

Sometimes, more than one model can explain the available data without discernible difference. In that case, one typically invokes *Occam's razor* to support the simplest explanation over the more convoluted (A related idea in model building is the *K.I.S.S principle*: Keep it Simple, Scholar!). However, all such support is tentative, until more exploration strengthens our case or causes us to revise our views.

We use current principles, models and theories to push the boundaries of knowledge and to make predictions that can be experimentally tested. If new, verified, observations do not support the existing framework, then the latter must be modified to provide an even better and more encompassing description of Nature.

Some distinguishing features of the scientific method are:
(1) *Falsifiability* of the hypothesis – that is, one should be able to test the hypothesis.
(2) *Reproducibility* of results — the same experiment conducted by independent examiners under the same conditions should give compatible results (in quantum mechanics the outcome need not be identical each time but must nevertheless follow the predicted statistical law).

2.2   Mathematical Models

Mathematics is a precise language, with an inbuilt logic, so it is ideally suited for the scientific study of patterns and trends.

Mathematical models usually start with several approximations and simplifications to capture the essential aspects of a phenomenon in a manageable form, and to allow for systematic improvements to the model. Of course, deciding what is vital in a model is an art!

For example, many early explorers modelled the Earth as a flat surface. That model was later replaced by that of a sphere, which was again improved to a squashed sphere (with the diameter in the North-South direction smaller than in the East-West direction).

Although we now know that the Earth is far from spherical, the simpler spherical model, or even the flat Earth model, is a useful approximation in some contexts.

In some situations, either for convenience or when a comprehensive theory is not yet available, two of more models might be used to describe different aspects of the same system[2].

A common error is to forget the limitations of a model and to use it where it is not valid; a related error is to confuse the model with the actual phenomenon that is being modelled. See, for example, Chap.(4).

Examples below, and in later chapters, illustrate some elementary aspects of the modelling process and the scientific method.

2.3 Examples of the Scientific Method

1. The ancient Greek philosophers tried to use their logical and argumentative abilities to summarize the workings of the world in terms of a few concepts. Sometimes this natural philosophy led to surprisingly accurate deductions, but more often than not the speculative reasoning led the natural philosophers to misleading or unhelpful conclusions: A dropped stone fell straight to the ground because 'that was its natural state', whereas a feather fell much slower and in an erratic manner 'because that too was its nature'.

   It would take many years before Galileo would argue that both the stone and feather would fall at the same rate, and straight down, once the wind and air-resistance were neglected. This was a crucial development because it illustrated that profound facts can be deduced if we focus on simple things. *Controlled*

---

[2]For example, it is sometimes convenient to think of light as a stream of particles, and at other times as a continuous wave.

experiments are done whereby only those factors one is interested in studying are allowed to vary while the rest are kept fixed.

Such controlled experiments are idealizations of natural phenomena and are meant to uncover the underlying rules of how the world works. The use of controlled experiments, and the scientific method of comparing hypotheses and theoretical predictions with experiments, has been the key to the success of the natural sciences.

By uncovering the few fundamental laws behind diverse phenomena, the world appeared to become more understandable and simple. With regard to the example above, Newton later showed that not only did the stone and feather fall to the earth at the same rate, so did the Moon: The law of gravitation is universal. This *universality* of fundamental laws is what makes us believe in the ultimate simplicity of the world.

2. When Mendeleev listed the known chemical elements in order of increasing atomic mass, he noticed that elements with similar chemical properties occurred at periodic intervals. This led him to his periodic table, which had several gaps filled by later discoveries.

   The understanding of the empirical patterns would come later once the nature of the atom, in particular the arrangement of electrons in orbitals and their role in chemical properties, was understood.

   This example illustrates the importance of recognising patterns in data during the early stages of scientific investigation.

3. Since the earliest times, we have perceived temporal regularities in Nature: Day to night and day again, the recurring seasons, the motion of the moon and other planets. Ptolemy held that the Earth was the immobile centre of the cosmos with the Sun and other planets revolving around it. Later Copernicus proposed a simpler scheme : The Sun at the centre with the planets revolving in circular orbits around it. The two proposals made different predictions that could be tested once accurate measuring instruments were invented to quantify the observations.

Later Kepler obtained results, such as the elliptical shape of the orbits, which improved on Copernicus' model but his laws were empirical, without any underlying explanation. That would come later with the work of Newton and his law of universal gravitation: Kepler's laws are simple consequences of Newton's more fundamental laws which also explain terrestrial phenomena.

However, even the great Newton had to bow to improvement when Einstein's General Theory of Relativity was verified by experiment and found to give a more accurate account of Nature.

This example illustrates how our understanding of the world typically improves and is refined as we get more information about it through technological progress — that in turn depends on our improved scientific understanding.

4. Around 1930, it was found that the neutron apparently decays according to the formula

$$\text{neutron} \rightarrow \text{proton} + \text{electron} \quad (??) \quad\quad (2.1)$$

Unfortunately, the above process violates energy conservation. One option considered by some physicists of that time was to give up their cherished principle of energy conservation!

However, another possibility was suggested by Pauli who postulated the existence of a new particle that he named the neutrino: A third particle on the right-hand-side of the above equation would allow the emitted electrons to have a range of velocities. In order to fit the experimental data, this particle had to be massless, and hence move at the speed of light.

The neutrino was indeed found by Reines and Cowan with the exact properties required to conserve energy.

This example illustrates the scientific method of observing phenomena, forming hypotheses, making deductions, and experimentation: Assuming some principles, such as energy conservation, one can make deductions (an undetected particle carrying away some energy) that can then be tested (experimental searches for the particle). See Fig.(2.1).

2.4   Common Misconceptions about Science

1. 'The scientific method only applies to the natural sciences such as physics, chemistry and biology.'
   The scientific method is a process of reasoning and is often used in daily life (see the exercises). While it is true that controlled experiments are harder to perform on human subjects, and modelling social systems is more intricate, yet the scientific method is still applicable as we will see in later chapters.

2. 'Science is the completely objective study of phenomena'.
   This cannot be true because the aspect of a phenomenon chosen for study depends on the scientist and is, therefore, subjective. Furthermore, the formation of a hypothesis or model depends on the individual scientist — the *art of science*.

   In other words, there is an important subjective, and creative, aspect of science, but it is constrained by the scientific method: The need for validating the theoretical construct against empirical data. In this way science achieves its essential objectivity.

3. 'Scientists do not make mistakes'.
   Being human, scientists have all the usual failings. Errors can happen due to oversight, or when there is a rush to publicise to gain priority or glory. Personal prejudices may also cloud the interpretation and selection of data.

   Even papers that are published in reputable peer-reviewed journals may turn out to be wrong, again due to oversight or other reasons (explicit fraud does occasionally occur).

   However, on the whole, sooner or later, independent critical analyses by other scientists detect and correct such aberrations: While individuals scientists might fail, the collective enterprise is able to progress[3].

4. 'Science constructs perfect theories'.
   This is never possible because we have only partial information at any one time. It is actually about constructing the best possible model (that is, accurate and economical) at that moment

---

[3]Compare this situation with the robustness of an ant colony in Sect.(6.2).

in time. The models improve as more information becomes available. So one is dealing with a sequence of approximations.

For example, from Newton's gravity to Einstein's gravity one has a generalisation that has been empirically verified. Some physicists have proposed further extensions, going beyond Einstein's theory: Among these models, some contradict the empirically confirmed predictions of Einstein's theory and so they they can be rejected. However, some of the proposed extensions are currently consistent with Einstein's theory (the predicted deviations are not measurable at present). In this case, one can appeal to Occam's Razor to distinguish among the competitors.

5. 'If a model is mathematical, it must be scientific'.
   No! *Mathematics* is simply a language. For the model to be scientifically validated, its predictions must match the observed facts. It is very easy to construct mathematical models that do not reflect reality.

   However, there is value in using approximate or crude mathematical models in initial stages for heuristic understanding, or as a preliminary step to a deeper investigation. See Chap.(7).

6. 'If there is no convincing theoretical explanation for a purported phenomenon, then the discipline studying that phenomenon cannot be a science.'

   Before attempting to provide a theoretical framework for a phenomenon, one should first ascertain the reality of the phenomenon. That is, there must be convincing data. In fact, the early stages of most sciences involve the collection and cataloguing of patterns.

   Sometimes the theoretical explanation of the data might be incorrect (for example, Kepler's attempt to explain the origin of his laws), but if the data is robust then the failure of a particular 'theory behind the data' does not invalidate the entire enterprise.

   Another example is the disease scurvy, which affected sailors on long voyages in the early days. It was soon realised that it was a nutritional deficiency which could be avoided if, for example, fresh citrus fruit was consumed. However, it took a long time to reach a rigorous theoretical understanding of the disease and its link to vitamin C.

What is problematic is when an intricate theoretical framework is constructed to support dubious data.

7. 'By analysing a phenomenon, science destroys its beauty'.
   'Beauty is in the eye of the beholder'. Scientists usually find that understanding a phenomenon enhances its beauty, whether it be a rainbow or the flight of migratory birds.

8. 'Science will ultimately encroach on all disciplines'.
   We do what we can to push back the boundaries; it is only then that we will know how far we can go. There might be some logical limits, and we will probably always have metaphysics.

2.5   Exercises

1. Your cell-phone screen has gone blank. You press the power button but nothing happens.

   (a) List the possible reasons for this phenomenon and discuss how the scientific method may be used to solve the problem. Consider also the role of Occam's razor in prioritising the various hypotheses.

   (b) Discuss another example from daily life that illustrates the scientific method.

   (c) Discuss an illustration of the scientific method from your favourite academic discipline.

2. You are asked to investigate the unknown relationship between an independent variable x, and a variable y that is believed to be dependent on x. You are given two pairs of data points, (x_1, y_1) and (x_2, y_2).

   (a) Could you determine the relationship $y(x)$?

   (b) If not, how many points would you need to model the relationship?

(c) Could you create more than one model even with the number of points you chose in (b)?

(d) How could you narrow down the possibilities and choose the 'most plausible' or 'realistic model' from among a few alternatives?

(e) What is the difference between *correlation* and *causation*? How could you differentiate between those two possibilities?

3. In medicine, a *placebo* refers to a neutral substance given to patients, instead of a drug, and without the patients being told of the actual contents of the pill (*blind tests*). It is sometimes found that patients given a placebo show as much improvement of their medical condition as those given the drug, leading to the suggestion that 'belief in the treatment, or positive feelings, have a healing effect on the body' (placebo effect). However, some critics point out that often, over time, the body naturally heals itself of many ailments, and thus one cannot conclude that the placebo effect is real.

(a) Design, with explanation, an experiment to study whether or not the 'placebo effect' is real, that is, whether placebos (in blind tests) can lead to healing.

(b) Are there any benefits to using the placebo effect in the treatment of medical conditions?

(c) Could the seeming efficacy of some popular 'alternative therapies' be due to the placebo effect? Support your claim.

4. Design a careful experiment (or a sequence of experiments) to test a new drug that claims to cure a particular illness. The drug has already been found safe in animal testing and the human testing has to begin. Clarify the role of the placebo effect, the importance of *double-blind testing*, the use of control groups, and the group sizes.

5. Dr. Hu published an article in a reputable peer-reviewed journal. In it he reported on a study that seems to show the ability of the herb Xanthinum to cure the common skin problem Dermimoxis. The study involved 159 individuals and another 144

in a control group who were given a neutral substance. The
reported results were analysed and found to be statistically significant.

Can we accept as a scientific fact that Xanthinum is a useful
remedy against Dermimoxis? Explain your position.

6. The government of Mooland, thinking of patenting a potentially
profitable discovery, has appointed you as chief scientific officer
to investigate the following claim made by some farmers: "Cows
produce more milk when they listen to soft soothing music".
Describe clearly how you would design an experiment to investigate the claim, and how you would take care of various factors
that might influence the results and their interpretation. Your
funding is limited to a reasonable amount and the maximum
time frame for the study is one year. See Ref.[2].

7. What is the difference between a sceptic and a cynic?
Is it possible to be sceptical of everything in the world and our
interactions with it?

8. Some scientists believe that all humans originated from a single
tribe in Africa. Discuss how this hypothesis has been investigated and what evidence supports it.

9. There appears to be a high correlation between the skin colour
of humans native to a region of Earth and the intensity of ul-
traviolet (UV) radiation at that location. Discuss a hypothesis
which might explain this, and how that hypothesis may be validated.

10. Explain, with examples, the differences between science, pseudo-
science, non-science, and nonsense.
Can you guess what these phrases might mean: 'Bad science',
'crackpot science', and 'proto-science'. Give explicit examples
to illustrate your meaning.

11. Debate one of these topics with your colleagues:

(a) Science and religion are ideologically incompatible: You
have to side with one or the other.

(b) There are many things we do not yet understand. Scientists should, therefore, keep an open mind to phenomena such as extra-sensory perception (ESP).

(c) Pick from one of the following and express your opinion on whether or not the activity is scientific: Astrology, feng-shui, numerology, palm reading.

(d) Pick from one of the following and express your opinion on whether or not the activity is scientific: Acupuncture, homoeopathy, traditional chinese medicine (TCM), Ayurveda.

(e) Scientists in the public sector (in universities and publicly funded research institutes) should work exclusively for the betterment of society, and also engage in their activity with the highest ethical and moral standards, as it is society which funds the research.

(f) Science is a social construct like all other forms of human activity; it cannot claim to describe an objective reality, especially since no one knows what reality is. Hence we should put no more faith in science than in other opinions and philosophies.

(g) The dominant reductionist approach used in Science, especially the physical sciences, will miss out on aspects of phenomena that are holistic, that is, which cannot be broken down into component parts or explanations. Indeed there are whole disciplines that must be studied holistically to form an accurate understanding (which will be impossible in a reductionist scheme).

2.6 References and Further Reading

1. The Sciences: An Integrated Approach,
   by J. Trefil and R. Hazen (Wiley, 2012).

2. http://www.le.ac.uk/press/press/moosicstudy.html

Notes

Chapter 3

Self-Organisation

The phrase *self-organisation* emphasizes 'organisation', or the creation of order, without directed external influence or deliberate internal effort.

Therefore, self-organisation may be defined as the spontaneous formation of macroscopic order (pattern formation) in a system whose components interact via relatively simple rules.

The idea of self-organisation was well established in the physical sciences before it was accepted in other disciplines. In mathematical biology, the basic idea was anticipated by Alan Turing but his work was appreciated only much later.

In the following sections and chapters, we will look at various examples of self-organisation.

We will also see in Chap.(4) that self-organisation is a form of *emergence*.

3.1 Swarm Intelligence

When an ant leaves its nest in search of food, it lays a chemical (pheromone) trail that enables it to find its way back home. If it does find food, it lays more pheromone on the return leg. That trail can then be used by the same ant or others as a guide to the food source [1]. The pheromone concentration on successful trails is thus reinforced through *positive feedback*.

In a classic experiment described in Ref.[2], ants were constrained by barriers to moving only along two paths from their nest to the food source. One path is direct and shorter compared to an alternative curved path, see Fig.(3.1).

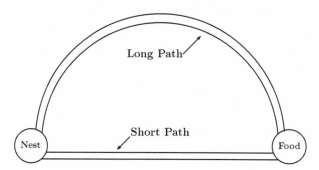

Figure 3.1: Ants search for food. A schematic drawing of the experimental layout mentioned in the text.

Some time after the start of the experiment, most of the ants were using the shorter path. The ants were able to find the shortest path to the food source, even though they have poor vision and are not able to judge distances directly. How did they do it?

Imagine initially five ants starting out on the direct path and another five simultaneously taking the curved path. Both parties will find the food source and return along their original paths, reinforcing their trails. However, those along the direct path will return earlier and be able to recruit other ants before those from the longer path. This positive feedback will continue to enhance the shorter path over the other and eventually most ants would go through the shorter path.

The best path found by the ants through their collective efforts is clearly an example of self-organisation. Similar *swarm intelligence* is exhibited by other social insects such as bees and termites [2, 3].

We will revisit ants in Chap.(6).

3.2   Spontaneous Symmetry Breaking

The concept of self-organisation is essentially equivalent to the concept of *spontaneous symmetry breaking* studied much earlier by physical scientists [4]. The 'self' corresponds to 'spontaneous', while 'organisation' refers to the creation of 'order' which implies 'symmetry-breaking'.

Why does order imply symmetry breaking?

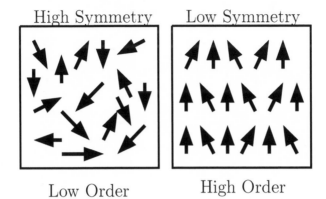

Figure 3.2: A system in two states. The left state has high symmetry and low order, while the right state has low symmetry and high order.

Look at Fig.(3.2) which shows a system of agents, represented by arrows, in two different states. Imagine yourself located at the centre of the system.

In the state labelled 'high symmetry', the system would look roughly the same in each direction from your central position. We say, the system has 'rotational symmetry'[1]. On the other hand, the state labelled 'low symmetry' clearly has a preferred direction and will not be rotationally symmetric from your perspective.

However, in terms of organisation, the 'low symmetry' state is clearly more ordered[2].

More concretely, if the figure represented a magnetic material with each arrow representing a tiny magnetic domain, then the ordered, low symmetry, state would display magnetism macroscopically.

On heating a magnet, the microscopic domains gain energy, become agitated and disordered, causing the microscopic state to have high symmetry while macroscopic magnetism is lost.

Conversely, if the non-magnetic material is cooled, the microscopic domains tend to line up, so the symmetry is spontaneously broken as the material becomes ordered and magnetic.

---

[1]That is, the word 'symmetry' is used here in its mathematical sense.

[2]This might sound counter-intuitive if you are used to loosely (and incorrectly) associating 'high symmetry' with 'high order'.

In passing, we note that spontaneous symmetry breaking is an essential ingredient in the Standard Model of particle physics, explaining how the various particles gain mass.

3.3   Organisation and Disorganisation

The slime mold is a fascinating organism that can exist in two distinct phases.

In one phase it consists of independent single cells. When they run out of accessible food, the individual mold cells release a chemical as a distress signal that attracts other cells to release more of the chemical, hence reinforcing the signal.

Interacting through the local chemical trail, the individual cells can self-organise into a new phase, that of a multi-cellular organism containing thousands of cells. The multi-cellular organism moves about in search of food and eventually grows to a shape consisting of a stalk and a cap containing spores [5].

The spores are released when the environment is ideal, germinating to form again a multitude of independent cells. The cycle of organisation and disorganisation repeats.

This example might have relevance for human organisations, see exercises in Chap.(6).

3.4   Pattern Formation

Consider a thin layer of liquid between two large parallel plates as shown in Fig.(3.3). If the system is in equilibrium, with the liquid and the two plates at the same temperature, and the liquid motionless, then the properties of the system are homogeneous.

If now the bottom plate is heated slowly, the heat will pass from the bottom plate to the liquid and will be transferred through the liquid to its upper layer by the process of thermal conduction. In thermal conduction there is no bulk motion of the liquid but rather a greater thermal motion of the molecules that causes the transfer of heat from the warmer layers to adjacent cooler layers.

However, as the temperature of the bottom layer is increased, a stage is reached where the liquid overcomes its internal friction and begins to undergo bulk motion. This results in a transport of energy by convection currents.

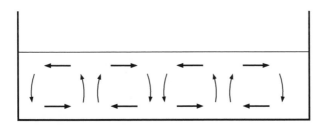

Figure 3.3: Benard cells, side view. The bottom plate is heated and is at a higher temperature than the top plate.

Initially, one typically sees small convection cells, called Benard[3] cells, as shown in the figure. This spontaneous coherent motion of molecules that were originally disordered is, therefore, another example of self-organisation [6, 7].

The size of Bernard cells is about 1 mm, which is much larger than the scale of intermolecular forces, 10^{-7} mm. The pattern formed by the cells breaks the original uniformity (translational symmetry) within the liquid [4].

This example illustrates clearly the open nature of self-organising systems. In this case, an inflow of energy in the form of heat results in non-equilibrium[4] structure formation on macroscopic scales [6]. By contrast, for the example of Sect.(3.2), order is attained at equilibrium by cooling the system.

The constraint placed by the Second Law of Thermodynamics, on increasing order in a system, is discussed in Chap.(4).

3.5 Temporal Order

Unlike most chemical reactions, where a state of homogeneity and equilibrium is quickly reached, the Belousov-Zhabotinski (BZ) reaction[5] is remarkable in maintaining a prolonged state of non-equilibrium [6].

---

[3]Named after the physicist H. Benard.
[4]There is a temperature gradient.
[5]Named after the chemists B. Belousov and A. Zhabotinsky.

The actual BZ reaction is very complicated, but its main characteristics may be illustrated using a simple model that the interested reader can find in Ref.[8]. Here we will content ourselves with some qualitative observations.

A crucial ingredient of the BZ system is the presence of *auto-catalytic* reactions. Recall that a catalyst is a substance that speeds up a chemical reaction, but itself remains unchanged; for example in $A + B + C \rightarrow D + C$, C is the catalyst. In an auto-catalytic reaction, the reactants produce more of the original catalyst, leading to a positive feedback loop, for example $A + B + C \rightarrow D + 2C$. The presence of auto-catalytic reactions means that rapid changes can take place.

If the reactants of the BZ system are well-stirred in a beaker, then for the right initial concentrations one observes oscillations in the system colour. This behaviour of the system, which can last a few minutes, is called a 'chemical clock' [6].

In this case the self-organisation, involving coherent action of the molecules, is temporal rather than spatial; see also exercise (7).

3.6   Patterns in Nature

Nature abounds with examples of beautiful patterns, such as those on sea-shells or butterflies. An attempt to understand such patterns from a physio-chemical perspective refers again to the BZ reaction [4].

If the ingredients of the BZ reaction are combined in a shallow dish without stirring, there will be some spatial differences in the concentration of reactants. Then, because of the auto-catalytic reactions, the differences will be amplified[6] leading to propagating wave fronts of different coloured chemicals [6].

A *Turing structure* is a stationary pattern formed when the propagating chemical waves of the BZ type are combined with the effects of diffusion in a medium (this is also referred to as a reaction-diffusion system) [4].

As expected for dynamic structures, the particular patterns that are formed in experiments and simulations depend on the initial conditions. This means that one can obtain high diversity. Furthermore,

---

[6]Chemical reaction rates depend on the concentration of the reactants.

the patterns depend on the geometry of the surface, and this is consistent with observations: For example, the beautiful patterns seen on ladybug beetles have been simulated in a model defined on a spherical surface, see Ref.[9].

In the 'Game of Life' simulation in the next chapter, we will see how diversity can arise from the sensitivity of a system to initial conditions.

3.7 Self-organised Criticality

There are many natural phenomena that exhibit *power laws*, of the form $N(E) \sim E^{-b}$, over a broad range of the parameters: Earthquake or volcanic activity, solar flares, length of streams in river networks, forest fires, and even the extinction rate of biological species [4].

Some of these power laws refer to spatial structures while others refer to temporal events. Can the frequent appearance of such power laws in complex systems be explained in a simple way?

The systems mentioned above are open and out of equilibrium, with a slow but constant inflow of energy and its eventual dissipation [6]. On the other hand, the above-mentioned systems display *scale-free*[7] behaviour similar to that exhibited by equilibrium systems near a *critical point* of a second-order phase transition (such as the transition point at which a magnet loses its magnetic properties when it is heated).

However, while the critical point in equilibrium systems is reached only for some specific value of an external parameter, such as temperature, for the non-equilibrium structures mentioned above the scale-free behaviour appears to be robust and does not seem to require any fine-tuning [10, 4].

Per Bak [10] and collaborators proposed that many complex systems self-organise to a critical state, with the consequent scale-free fluctuations giving rise to novel (emergent) properties, in particular power laws.

Their proposal is that self-organised criticality (SOC) is the natural state of large complex dissipative systems, relatively indepen-

---

[7]'Scale-free' means the distribution does not have a peak representing a 'typical' value or scale. See also 'scale-free' networks in Chap.(5), and the chapter on Fractals.

dent of initial conditions. While the critical state of an equilibrium second-order phase transition is unstable (slight perturbations move the system away from it), the critical state of self-organised systems is stable: Systems are continually attracted to it.

In addition to the examples mentioned above, the self-organised criticality idea has also been proposed to apply to economics, traffic jams, and the brain [10, 4].

3.8   Summary

We looked at some real-world examples of spatial and temporal self-organisation. Self-organisation does not contradict the Second Law of Thermodynamics, see Sect.(4.7.1).

In the next chapter, we will look at simple computer simulations that illustrate self-organisation and its relation to the concept of emergence.

3.9   Exercises

1. Can a system be

   (a) 'Self-disorganised'?

   (b) Organised but not self-organised?

   (c) Partially self-organised and partially planned?

   Illustrate with examples.

2. Which of these are examples of self-organisation:

   (a) Crystallisation (of a chemical in a beaker in a laboratory).

   (b) Formation of natural snow-flakes.

   (c) The generation of a laser beam.

   (d) Wikipedia.

   (e) Biological evolution through natural selection.

   (f) The formation of our Solar System.

   (g) The propagating wave formed when a stone is thrown into a pond.

Explain.

3. Using the example of an ant colony searching for the shortest path to a food source,

   (a) Explain the terms 'positive feedback' and 'negative feedback'. (Hint: The pheromone evaporates).

   (b) Why do you think some ants still take the longer path even after most ants have converged to the shorter path? Could it be due to over-crowding?

   (c) What advantage could there be to some ants not following[8] the crowd as mentioned in part(b)? See also Sect.(6.2).

4. Social insects such as bees have relatively simple individual behaviour but they seem to be able to perform some complex tasks. Explain clearly how such simple agents, following simple rules, can accomplish such tasks. Is it important for there to be many of the simple agents? Explain.

5. Do Benard cells occur in Nature?

6. Read about the Turing mechanism in Ref.[4].

   (a) Discuss one main advantage in using the Turing-structure explanation for observed animal skin patterns as opposed to an alternative hypothesis relying on a detailed blueprint coded by the DNA.

   (b) Look for potential Turing structures in Nature and discuss their similarities, differences, and possible functions or advantages from the perspective of evolution and natural selection.

   (c) How can geometry affect the final Turing pattern? Explain with an explicit example.

   (d) Are you convinced that the Turing mechanism is responsible for animal skin patterns? Explain.

---

[8]In some human organisations such agents would be labelled as 'not team players'.

7. It has been shown that in certain species of firefly, large populations of them can synchronise their flashing simply by keeping time and observing the flashes of their nearest neighbours [11]. It has been suggested that this is an example of 'self-organisation'. Explain clearly, with respect to this example, the concept of self-organisation.

8. Heat can drive a system to order or disorder, as seen in two examples in this chapter.

   (a) Can you identify the other factors which determine the likely outcome?

   (b) Is symmetry broken when Benard cells form starting from an initial homogeneous state?

9. Your friend Curioso claims: "Looking for simple rules between the building blocks of a complex system cannot explain, and in fact it will miss out on, holistic aspects of the system."
   Do you agree with that claim? Give a clear and convincing example to support or counter the above claim.

10. Simplicio criticises the Turing structure models of pattern formation in biology: "The differential equations are not directly related to the underlying biochemistry and furthermore the parameters in the equations are chosen precisely to obtain chosen patterns. The whole effort is simply an *ad hoc* exercise and has no predictive power, nor is it falsifiable". Comment.

11. What determines the direction of spontaneous symmetry breaking during the cooling of a demagnetised material?

12. What evidence is there that some phenomena are well explained by the concept of self-organised criticality?

3.10 References and Further Reading

1. The Ants, by B. Holldobler and E. Wilson
   (Belknap Press, 1990);
   Ants at Work, by D. Gordon (Free Press, 1999).

2. 'Self-organized short-cuts in the Argentine ant',
   by S. Goss et. al, Naturwissenschaften 76 (12) (1989), 579.

3. 'Swarm Smarts', by E. Bonabeau and G. Theraulaz,
   Scientific American, March 2000, 72.

4. The Self-Made Tapestry, by P. Ball
   (Oxford University Press, 1999).

5. Signs of Life, by R. Sole and B. Goodwin (Basic Books, 2002).

6. From Being to Becoming, by I. Prigogine
   (W. H. Freeman, 1980).

7. The Web of Life, by F. Capra (Anchor Books, 1996).

8. Designing The Molecular World, by P. Ball
   (Princeton University Press, 1996).

9. 'Turing model for patterns of lady beetles',
   by S. Liaw et.al, Phys. Rev. E64 (2001) 041909.

10. How Nature Works, by P. Bak (Copernicus, 1999).

11. Sync, by S. Strogatz (Hyperion, 2012).

Notes

Chapter 4

Agent-Based Models and Emergence

In *agent-based modelling*, one creates a model in which each component of the system is treated as an *agent* that interacts in a well-defined way with other agents or the environment.

When multiple agents interact, one often observes behaviour or a pattern at the system level that is not an intrinsic property of any individual agent or an obvious consequence of the rules: Such behaviour or pattern is termed *emergent*[1].

From our definition of 'emergence' and the previous definition of 'self-organisation', it is clear that all self-organised phenomena are emergent. However, as we will see, the converse need not be true.

4.1 Termites

We begin with an example in which self-organisation is manifested even when the agents do not directly interact with each other. Consider a termite-like agent following the following three simple rules in the model created by Resnick [1]:

1. It moves via a *random walk* until it collides with a wood chip.

---

[1]This will be our working definition of 'emergence', which might differ from other uses of the term in the literature. The less intuitive word 'epiphenomenon' is sometimes used instead of 'emergent'.

2. If the termite is not carrying a wood chip, it picks up the one it bumps into and continues its random walk.

3. If the termite is carrying a wood chip when it bumps into another, it drops the chip and continues its random walk.

A *random walk* is a path generated by a random process: Consider a two-dimensional random walk starting at the origin. The location of the the next step is generated, for example, by two random numbers which give respectively the direction (angle) and length of the walk. The random process is repeated at the following time intervals.

Although the movement of the termites is random, the rules for picking or dropping the wood chips are completely *deterministic* — that is, precisely specified, with no randomness involved. A simulation, for example [2], shows that a small number of termites following those rules spontaneously rearranges an initial random configuration of wood chips into clusters.

Though acting independently and locally, with no global plan, nor a leader, nor any external direction, the termites' effort gives rise to order. The self-organisation in this case can be made more realistic by changing the rules slightly – see Refs.[1, 3].

It should be borne in mind that self-organisation need not be inevitable: It might depend on certain variables in the model, for example the number of agents and the amount of wood-chips.

4.2   Proof of Concept

In the previous chapter, we illustrated the concept of self-organisation through several real-world examples. However, a critic might justifiably demand proof that self-organisation occurred in those examples — could it be that we have simply missed some directed external influence, or an internal 'leader', that might explain how the organisation occurred?

It is, therefore, useful to demonstrate the *proof of concept*: We do this by creating simple models, with precisely specified rules, that display the idea we wish to illustrate.

Resnick's termite model provides proof for the 'self-organisation' concept. Although the model is far from any realistic portrayal of termite colonies, it serves its purpose by illustrating the self-organisation idea unambiguously.

Several other examples in this and other chapters may be viewed as 'proofs of concept', if not as simplified models of actual phenomena.

4.3 Boids

One of the earliest examples of interacting agents is Craig Reynold's 'boids' model [4, 5] that simulates the motion of a flock of birds. In this model, each boid is an agent that follows a few intuitive rules meant to optimise some objectives:

1. Separation: Reduce the chance of collisions by moving away from boids that are too close.

2. Alignment: Fly in the average direction that the flock is moving.

3. Cohesion: Move towards the centre of the flock to avoid exposure at the exterior.

Reynold's rules are illustrated in the figure below.

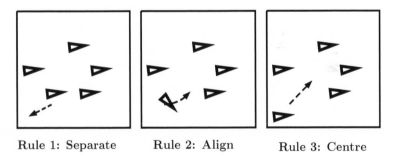

Rule 1: Separate   Rule 2: Align   Rule 3: Centre

Figure 4.1: Boid rules.

A simulation of the three rules (you can check out such simulations on the WWW) shows that an initially disorganised collection of boids comes together to form a single flock flying coherently.

(The position of a boid after each increment in time step, assumed here to be of unit length, is obtained by adding to the old position the new weighted velocity that takes into account the above rules. Please see Refs.[4, 3] for details.)

4.4   Models versus Reality

Birds that fly long distances have been observed to adopt a 'V' shaped formation. To achieve such a pattern within the simple boid model, Flake [3] added the following fourth rule:

"4. View: Move laterally away from any boid that blocks the view".

Remarkably, the four rules together do give rise to the classic V-formation [3]. However, this does not imply that real birds behave according to those rules.

Studies have shown that birds experience less resistance in their flight, thus minimising energy usage, when in a V-shaped flying pattern. So, the actual rule should probably be 'move sideways to conserve energy' — but such a real-world rule would be very difficult to implement in a simple computer model, such as the boid model, which does not take into account the physical properties of the medium.

This example reminds us: Models that replicate observed patterns need not be entirely consistent with the facts.

In other words, it is always possible to create different models to explain the same limited amount of data. The competing models can be distinguished from each other by their ability to explain or predict new data not already incorporated into the model.

4.5   Game of Life

The mathematician John Conway invented a game in 1970 that shows some life-like features [6].

The game is run as a cellular automaton in discrete time-steps. A *cellular automaton* is a special type of agent-based simulation: It is run on a two-dimensional lattice of cells, each of which interacts with its eight neighbours via a few simple rules. See Fig.(4.2).

In the Game of Life, the cells of the lattice take one of two states: Alive (dark square) or dead/vacant (white square). At each time-step the state of a cell is determined according to the following rules:

1. A living cell will remain alive only if it has two or three living neighbours. It dies from loneliness if it has less than two

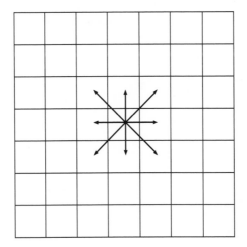

Figure 4.2: Neighbours of a cell in a cellular automaton.

neighbours, and from overcrowding if it has more than three neighbours.

2. A dead (vacant) cell can come alive if it is surrounded by exactly three live cells. One can think of this as reproduction.

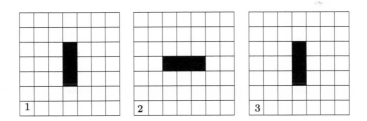

Figure 4.3: A periodic configuration shown at three successive time steps: Three horizontal live cells become a vertical line, and then turn back to the original configuration. The process obviously repeats.

Though the above rules are a caricature of ecosystems, they lead to surprising and intricate patterns. Indeed, Conway's game is a beautiful exemplification of emergent behaviour (see the applets on WWW).

Starting with different configurations of living cells, many possible outcomes can result, including periodic patterns, gliders and breeders, see Fig.(4.3) and Fig.(4.4).

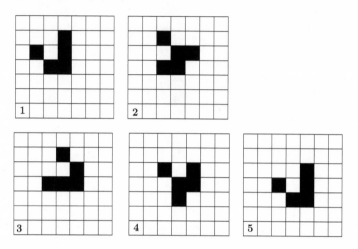

Figure 4.4: A glider shown at five consecutive time steps: The configuration at the top left glides to the right, changing shape slightly at intermediate stages.

4.6　Lessons from the Game of Life

The Game of Life simulation illustrates some important ideas in the general study of complex systems:

1. The observation of complex patterns does not imply that the rules of interaction between the agents of the system must be complex.

2. The same rules (universal rules) can lead to diversity simply by varying the initial conditions (arrangement and number of cells). In other words, in a good model, it is not necessary to keep changing the 'rules of the game' to explain the variety of observed patterns.

3. Even though the rules might be simple, this does not imply that the result will be easy to predict. (Try to guess what would happen if you started with four live adjacent cells in a row).

4.7 Thermodynamics

Thermodynamics is the study of the macroscopic physical properties of a large collection of atoms or molecules. That is, instead
of attempting to describe the individual motion of the microscopic
particles, one tries to make statements about the properties of the
system as a whole. One says that a system of atoms has reached
thermodynamic equilibrium when its macroscopic properties, such as
temperature, do not change with time.

The thermodynamic relations, such as the ideal gas law[2], $PV \propto T$,
are often good approximations when one is talking about the average properties of a system with a large number of atoms. Indeed,
the various thermodynamic relations are examples of *emergent laws* :
Generalities about the system that are apparent only at the macroscopic scale but are not obvious, or existent, at the microscopic level.

Even the concept of temperature is a macroscopic emergent feature that is ill-defined for a system with only a few atoms. The absolute temperature T of a gas, measured in Kelvins, is proportional
to the mean kinetic energy of the atoms or molecules making up the
gas.

4.7.1 The Second Law of Thermodynamics

The *Second Law of Thermodynamics* states that the entropy of a
thermally isolated system (in equilibrium) never decreases. *Entropy*
is a technical measure of the amount of disorder in a system.

To see how the above terms correspond to our intuitive use of
them, consider the following example: An isolated box contains an
equal number of two types of molecules, say 'white' and 'black' [8].
The molecules are in constant motion, colliding with each other and
with the walls of the box.

Suppose that, at a particular moment in time, we manage to seg-
regate all the white molecules to the right side of the box and the
black molecules to the left. Of course, this is an unnatural situation
and soon, due to their motion and collisions, the molecules will mix.

The initial situation, with segregated molecules, clearly corresponds to a state of maximum order of the *microstate* (the specifica-

---

[2]P is the pressure, V the volume and T the absolute temperature.

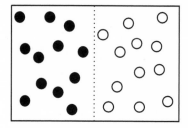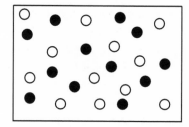

Figure 4.5: Molecules in a box.

tion of the location of the molecules), while the final mixed configuration corresponds to large disorder (randomness) of the microstate.

One can estimate the entropy of the ordered and disordered states: Clearly there are many more ways to form the disordered state than the ordered state and thus (from the definition of entropy) the entropy of the disordered state is much higher than the entropy of the ordered state. Since we know that the gases will mix, the system goes from the ordered state to the disordered state, increasing its entropy in accordance with the Second Law [8].

We also know that if the number of molecules is very large, it is extremely unlikely that the gases will revert to the totally separated state at some future time. Thus the increase in entropy and disorder in the system is, for all practical purposes, irreversible.

Note that the irreversibility is not due to the underlying physical laws, but the result of the system going from an unlikely ordered state to a more probable disordered state; and the fact that for large systems (large number of molecules) the probability of the system reverting to the ordered state being negligible [8].

In the above example, the probability of any one molecule being on the left half of the box is $1/2$. If there are N molecules, the probability that all of them are on the left is $(1/2)^N$. Even for N as small as 100 this works out to be about 10^{-30}, an infinitesimal quantity: For macroscopic materials, N is of the order of 10^{23}, and so the actual probability is much lower[3].

Thus, the Second Law is a statement about average behaviour that becomes overwhelmingly likely in a very large system, meaning

---

[3]Avogadro's constant, $N_A = 6.022 \times 10^{23}$, is the number of atoms in 12 grammes of Carbon-12 or 18 grammes of water.

that exceptions will be unobservable in all practical situations.

Since the individual molecules themselves are not constrained by the Second Law, which applies only to a large collection of the molecules, the Second Law is, therefore, an example of an *emergent law*.

4.7.2 Life

Living systems tend in their development towards greater order, in contrast to the arrow of time dictated by the Second Law of Thermodynamics.

Of course, there is no conflict as the increase in disorder and entropy required by the Second Law refers to closed equilibrium systems. Living systems are neither closed nor in equilibrium, but rather use an inflow of energy to drive processes that increase their order (thus decreasing their entropy), and dissipate heat and other waste products that lead to an overall increase in entropy of the universe.

4.8   Emergence versus Self-Organisation

We mentioned at the start of this chapter that, with our definitions, all self-organised phenomena are examples or emergence.

However, as the case of the Second Law of Thermodynamics shows, not every example of emergence is an illustration of self-organisation; indeed in this particular example the emergent law emphasises the move towards greater disorder.

4.9   Do it Yourself

Many researchers have created user-friendly software for the investigation of complex systems using agent-based models.

One such example is the NetLogo [9] tool-kit, which contains numerous models with adjustable parameters. It also enables the creation of new models with relative ease, requiring minimal technical pre-requisites.

Try it out.

4.10   Summary

Computer simulations show that agents obeying simple local rules can give rise to self-organisation or emergent properties, providing proof of those concepts.

Although the simulations are valid examples of emergence, one must nevertheless be cautious in interpreting the results. Agreeable results may make a model a plausible abstraction of a real system, but this does not rule out other models, though the simplest explanation is both appealing and fruitful for further investigation.

4.11   Exercises

1. Read the article on the Game of Life at Ref.[7].

   (a) Run the applet at Ref.[7] using the A-plus starting configuration. Identify some examples of static, periodic, moving, and breeding configurations. Convince yourself that the figures conform to the rules of the Game.

   (b) Try other starting configurations and see if you can find life-like, self-replicating structures.

   (c) How does diversity (different emergent patterns) arise in the Game of Life if the rules are so simple?

   (d) Explain in what sense the pattern formation in the Game of Life differs from real life: That is, what characteristics of living organisms are missing from those artificial creations?

   (e) How would you modify the rules to address the shortcomings identified in the last part?

2. A starting configuration in the Game of life has four adjacent live cells in a row.

   (a) Guess what the final configuration would be. Check using the applet.

   (b) Repeat the exercise of part (a) using 5 or 6 adjacent live cells in a row.

   (c) What can you conclude from these exercises?

3. Run the termite simulation at Ref.[2].

(a) Is there a stable conclusion?

(b) Does the result depend on the relative amounts of termites to wood initially? (You can change the initial amounts in the control box).

(c) Discuss the lessons to be learnt from this simulation.

4. Reynold's boid model.

(a) Try to justify the rules from an evolutionary perspective.

(b) Are there other animals that could plausibly obey similar rules for herding/flocking/schooling?

(c) Could there be other explanations for the formation of real flocks?

(d) Explain clearly with the help of a diagram what might result if the alignment rule is abandoned and only the other two rules implemented in the model.

5. Explore some of the herding simulation applets on the WWW. In what sense do those models exhibit self-organisation and emergence?

6. What is the difference between 'self-organisation' and 'emergence'?

(a) Think of an example of emergence or self-organisation that is not mentioned in the text. Justify your choice.

(b) Does your example illustrate both or just one of the concepts?

(c) Can an organised system display emergent properties that are not self-organised? If yes, give an example.

7. Read the section of Johnson's book [10] on DNA and the development of the embryo.

(a) Discuss the *top-down* versus *bottom-up* processes during embryo development.

(b) In what sense does this example illustrate self-organisation or emergence?

8. Can you tell whether you are living in a 'real' world or are just part of a massive agent-based simulation? Prove your assertion[4].

9. Which of these phenomena violate the Second Law of Thermodynamics?

    (a) Water frozen in a refrigerator turns to ice in which the motion of the molecules clearly decreases, and thus the the amount of disorder (entropy) of the system (water) decreases.

    (b) Patterns in the Game of Life, such as oscillators and breeders, not becoming disorganised over time.

    (c) The BZ chemical clock exhibiting coherent oscillations.

    Explain.

10. A student exclaims: "The second law of thermodynamics is a law of physics, not biology. In fact, living systems clearly violate that physical law!" Comment critically on the student's opinion.

11. Newton's deterministic laws of mechanics, which apply to macroscopic objects, are different from quantum rules which apply to the object's constituent atoms. Newton's laws are examples of emergent laws.
    List some other examples of emergent laws, explaining clearly why you believe them to be emergent. Are your laws universal? Explain.

12. Which of these is emergent? Explain.

    (a) The size of Bernard cells; see Chap.(3).

    (b) The smooth surface of water in a pond.

    (c) The new colour that results when two other colours are combined.

13. W. Weaver [11] suggested that 'complexity' may be viewed as being of two different types: 'organised complexity' and 'disorganised complexity'.

---

[4]What is real, really?

(a) Classify the examples from this chapter into those two types.

(b) Can you think of examples that are not easily classified into either of those types?

14. Use NetLogo [9] or other software to re-create one or more of the agent-based models discussed in this chapter. Verify the patterns. Experiment by modifying the rules.

4.12 References and Further Reading

1. Turtles, termites, and traffic jams, by M. Resnick (Bradford Books, 1994).

2. A termite simulation at
   http://www.permutationcity.co.uk/alife/termites.html

3. The Computational Beauty of Nature, by G. Flake (MIT Press, 1998).

4. 'Flocks, herds and schools', by C. Reynolds, Comp. Graph. 21(4) (1987) 25.

5. Craig Reynolds' Boids at http://www.red3d.com/cwr/boids/

6. Conway's Game of Life, an article by M. Gardner, Scientific American 223 (1970) 120.

7. A Game of Life simulation at
   http://www.math.com/students/wonders/life/life.html

8. The Feynman Lectures on Physics, Vol. 1, by R. Feynman, R. Leighton and M. Sands (Addison Wesley, 1977).

9. NetLogo at https://ccl.northwestern.edu/netlogo/

10. Emergence, by S. Johnson (Simon and Schuster, 2012).

11. 'Science and Complexity' by W. Weaver, American Scientist 36 (4) (1948) 536.

Notes

Chapter 5

Networks

Consider Fig.(5.1). Suppose each *node* in the figure represents a person (agent) in a certain community, and a *link* between two nodes indicates a friendship between them. The collection of nodes and links then represents a particular *network*, the 'friendship network' for that community.

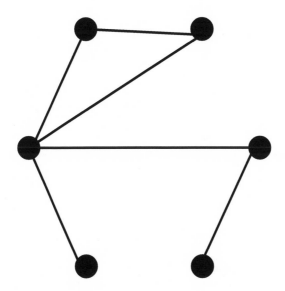

Figure 5.1: A network.

Other examples of networks are the metabolic network of cells, the world-wide-web (WWW) whose nodes are HTML[1] documents with links pointing from one node to another, the collaboration diagrams of scientists or film stars, and transportation networks [1].

Realistic networks often evolve, with the number of nodes and the links among them changing with time. In general, links can be directional (for example, as in an acquaintance network) but in this chapter we will consider mainly networks with non-directional links.

We will be concerned primarily with networks associated with complex systems, and refer to such networks as *complex networks*. Several questions can be asked about complex networks: How connected is the network? What is the shortest path from one node to another? What mechanisms determine the structure of a network? Are there any emergent features? How robust is the network?

In the following sections, we introduce the basic terminology and models in the study of complex networks. For details and applications, such as how processes (spread of information or viruses) take place on a given network, the reader is referred to the references, for example Ref.[2].

5.1   Properties

For any network, one can make measurements of at least three quantities of interest [2],
(1) The *average path length* between two nodes,
(2) The *clustering coefficient*, and
(3) The *degree distribution* of the nodes.

The path length d_{ij} between two nodes i and j is the number of edges along the shortest path connecting those two nodes[2]. For a fixed node i, averaging d_{ij} over j gives the average path length starting from node i. Finally, averaging over both i and j gives the average path length for the network.

The 'six degrees of separation' concept popularised by Milgram, and the more general 'its a small world' idea, implies that many social networks have a small average path length [3, 4].

---

[1] Hyper Text Markup Language.
[2] For networks where the physical distance between nodes is important, a 'weight' can be assigned to each link, see exercises.

A *completely connected* network is one in which every node is connected directly to every other node. An example is Fig.(5.2).

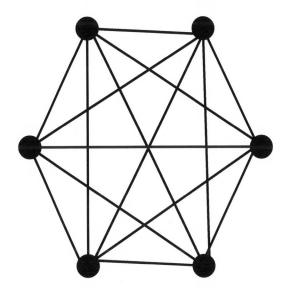

Figure 5.2: A completely connected network.

In a typical network, not all nodes are connected directly: Clustering measures the tendency of some nodes to be completely interconnected [3, 4]. In social networks, it would mean groups of individuals who all know each other. Let a node i have k_i edges connecting it to that many other nodes. If those k_i nodes were part of a completely connected sub-graph, there would be $k_i(k_i - 1)/2$ edges among them. Let E_i be the actual number of edges present. Then the clustering coefficient of that node is

$$C_i = \frac{2E_i}{k_i(k_i - 1)} \; . \tag{5.1}$$

Averaging the C_i over the whole network gives the clustering coefficient C for the network [3].

Finally, the degree distribution of the network, $P(k)$, quantifies the probability that a randomly selected node has exactly k edges.

5.2   Random Networks

Early studies assumed that large networks would most likely be random in their topology [2]. A *random network* can be constructed as follows: Start with N nodes and connect every pair with probability p. This will result in a network with approximately $pN(N-1)/2$ links distributed randomly.

It has been shown that random networks are indeed 'small worlds', with the average path length scaling as $\log(N)$.

For random networks, one has $C = p$. This is a much lower value than that observed in many real networks of the same size.

For large random graphs, the degree distribution is Poisson with mean (average) pN. That is, the distribution is peaked near the mean pN, so most nodes have degrees close to that value [1], see Fig.(5.3).

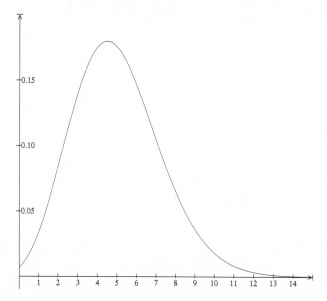

Figure 5.3: A Poisson distribution. The tail falls off exponentially.

However, many real networks have been found to have a very different degree distribution, a power law $P(k) \approx k^{-\gamma}$ tail being quite common[3].

---

[3]However, not all real-world networks have a power law tail [1].

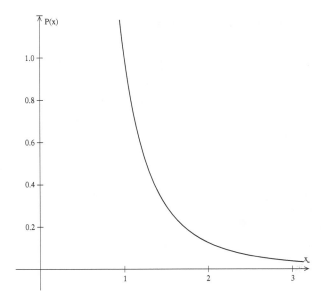

Figure 5.4: A distribution with a power law tail.

To summarise, there are two crucial differences between many real networks and models based on random networks. Firstly, real networks show a much greater degree of clustering than random networks.

Secondly, the degree distribution of real networks often displays a power law tail, meaning that there is no peak or typical value: Nodes with very large degrees, called *hubs*, are not uncommon. Specifically, hubs in real networks are often not exponentially suppressed as in random networks [4].

Networks with a dominant power law degree distribution are also called *scale free networks* as they do not have a peak representing a 'typical' value (scale).

5.3   Small World Networks

Watts and Strogatz [3] proposed a simple model that displayed both a short path length and large clustering, features seen in many real networks. Recall that random networks have short path lengths but low clustering. At the other extreme, completely connected networks, such as in Fig.(5.2), have high clustering but large path lengths.

Watts and Strogatz (WS) generated an intermediate type of network by the following algorithm:
(1) Start with an ordered network[4], and then
(2) Randomly rewire each edge with probability p.

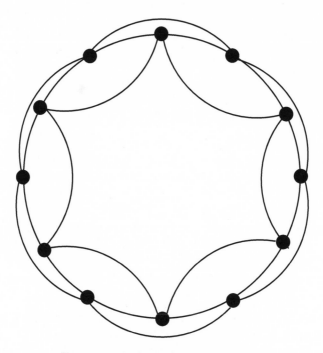

Figure 5.5: An ordered network.

The rewiring introduces some short-cuts between otherwise long separated nodes thus making the path length short; compare Figs.(5.5, 5.6). It was shown that in a one-dimensional ring lattice, high clustering emerged for some range of values of p [3].

Unfortunately, the WS model produces networks with a degree distribution similar to that of a random network, peaked at some value, unlike many real-world networks which appear scale-free.

---

[4]Each node in an ordered network has the same number of links.

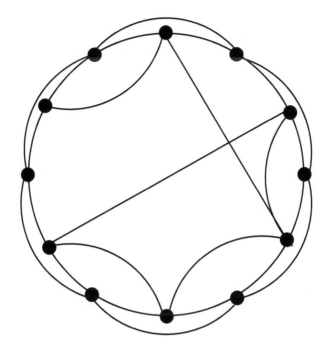

Figure 5.6: A Small World Network.

5.4   Scale-Free Networks

Barabasi and Albert [5] proposed a model that would generate a power law degree distribution. Their algorithm involves a growing network with new links formed by preferential attachment:

(1) Growth: Start with a small number of nodes and at every time step add a new node that links to m different nodes in the system.

(2) The probability $\Pi(k_i)$ that the new node will attach to a node i is given by $\Pi(k_i) = \dfrac{k_i}{\sum_j k_j}$.

That is, nodes with many links collect more links as the network grows —'the rich get richer phenomenon'. Numerical simulations [5] of the BA model show that the above procedure generates a power law degree distribution with exponent $\gamma = 3$. If either of the above two conditions is omitted, power law behaviour is not seen. To get other values for the exponents, and closer agreement with real network results, condition (2) can be generalised [2].

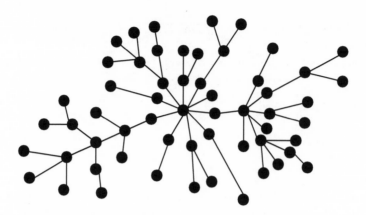

Figure 5.7: Growth of a scale-free network.

The power law distribution in the BA model is clearly an emergent feature.

What about the path length and clustering properties of the BA model? Fortunately, the path length is small (smaller than random graphs) while the clustering coefficient is several times larger than random graphs. Thus, the BA model provides a reasonable theoretical understanding of many evolving real-world networks, and it has been generalized and studied from various perspectives, including the modelling of dynamical systems (such as the spread of epidemics) on networks.

5.5   Network Stability

If random errors damage some nodes in a network, then the links connecting those damaged nodes to other nodes are also put out of action. Thus, one can ask about the size of the largest connected cluster that remains as some nodes are damaged [4].

For a random network, it is found that as the fraction of randomly removed (damaged) nodes is increased, the size of the largest connected cluster decreases to zero. The same situation is true for scale-free networks but now nearly all nodes must be removed for total failure. This is because there are some highly connected nodes (the hubs) in a scale-free network, and if nodes are removed randomly it is unlikely that most of the hubs will be eliminated early on.

On the other hand, a deliberate attack on networks targeting the highly connected nodes leads to the failure of the scale-free networks much earlier than the failure of random networks.

In summary, scale-free networks are robust to random errors but quite susceptible to intentional attacks on the hubs [4, 2].

5.6 Summary

The topology of many real-world networks, such as the WWW and metabolic networks, evolved spontaneously rather than being planned. Researchers have developed models of such networks to reflect the observed characteristics. Some of the key features are emergent.

5.7 Exercises

1. Draw a network representing members of your class, office or organisation.

   (a) Calculate the average path length of the network with respect to yourself (your node) as the fixed starting point. That is, calculate the average distance between you and every other person. (To do this systematically you might want to draw a tree diagram).

   (b) Identify the top four hubs of your network.

   (c) Redo part (a) but now with the top hub removed (that is, assume the hub and its directly associated links are removed. If the network becomes disconnected, take note of that and compute your average path length in the part of the network that you remain connected to).

2. Prove that if any complex network has a clustering coefficient equal to 1, then its average path length must also be 1. Sketch, with motivation, the degree distribution of such a network.

3. A network has the degree distribution $P(k = 2) = 1$.

   (a) Sketch the network.

   (b) Deduce its clustering coefficient.

(c) Determine the average path length (you may assume the number of nodes is large and so make some approximations).

(d) If a single new link is to be added to the above network so as to reduce the path length significantly, decide where you will place it. Calculate the new average path length and explain why the answer is reasonable.

(e) Discuss how adding the new link in (d) affects the robustness of the network to random failure of the nodes (assuming it is a communication network).

4. Let $k(i)$ denote the number of links connected to node i in a network. Prove that the sum of $k(i)$ over all the nodes equals $2L$, where L is the number of links in the network.

5. A student claims that his friendship network is ordered, with 7 nodes and 5 links per node. Is such a network mathematically possible? Explain. (Note: Links connect different nodes).

6. Let k_m be the largest number of links connected to a single node in a network. Prove that the network must have at least $k_m + 1$ nodes.

7. Draw the smallest networks that have the degree distribution $P(k = 3) = 1$ and

   (a) $C = 1$ (C is the clustering coefficient).
   (b) $C \neq 1$.
   (c) In each of the above cases discuss the robustness of the networks to random failure of a few (one or two) nodes.

8. Sketch the smallest (connected) network which has the degree distribution $P(k = 2) = P(k = 3) = 1/2$.

   (a) Compute the average path length of your network.
   (b) Compute the clustering coefficient of your network.

9. A connected network consists of 6 nodes lying at the vertices of a hexagon. The nodes are labelled 1 to 6 clockwise. Six of the links form the sides of the hexagon. Two other links connect vertices (2,6) and (3,5) respectively.

(a) Sketch the network.

(b) Compute the average path length of the network.

(c) Compute the clustering coefficient of the network.

(d) Determine the degree distribution of the network.

10. The Barabasi-Albert (BA) model.

    (a) How plausible are the BA rules as mechanisms operating
        in real networks? Discuss in the context of two explicit
        examples.

    (b) How can the BA rules be modified to generate other expo-
        nents for the power law tail?

11. Are there real networks with a degree distribution whose tail is
    not a power law?

    (a) What could be the operating mechanism or cause for the
        non-power law behaviour?

    (b) Discuss the robustness and susceptibility of one such real
        network.

12. The government of Stopia has decided to link its departments
    (nodes) with optical fibre cables for secure communications.
    The criteria are as follows:
    (i) The network should be robust against intentional attacks on
    any of the nodes, that is, the network should not become dis-
    connected too easily,
    (ii) Messages should not have to pass through too many inter-
    mediate nodes before reaching their destination node, and
    (iii) In order to remain cost-effective, the network should not
    have too many links.

    As a complex networks expert, you have been asked to propose
    a topology for the network, that is, how the nodes are to be
    linked to one another. Illustrate your proposal with the help
    of a diagram for the case of eight nodes lying uniformly on the
    circumference of an imaginary circle. Explain clearly how your
    proposed topology would achieve the specified criteria.

13. A communication network consists of $N + 1$ nodes connected in a 'wheel formation' as follows: N of the nodes are arranged uniformly around the circumference of a circle while one node is at the center of the circle. Every node on the circumference is connected directly to a node at its immediate left and to a node at its immediate right on the circumference. The central node is connected directly to each of the circumference nodes. (Thus each circumference node is connected directly to three nodes, while the center node is connected directly to N nodes).

   (a) If each radial link is equivalent to 2 circumference links, determine the average path length l of the network for very large $N(\gg 1000)$. Show your steps and reasoning.

   (b) A deliberate attack is launched on that network to destroy one node and hence disrupt all links connected through it. Estimate, with an explanation, the new average path length after a worst case scenario.

   (c) How would you re-design the original network so that the average path length does not change dramatically after a deliberate attack on a single node. Your new network should still have $N+1$ nodes but have 1/3 fewer links than the original network. Justify your solution and provide a clear sketch of your proposed network.

   (d) What is the clustering coefficient for each of the networks in parts (a,b,c) above? Provide clear reasoning.

14. It has been suggested that both biological networks and human-engineered networks share the structural principle of 'modularity', see Ref.[6].

   (a) Explain what 'modularity' means by using an example network from either biology or human engineering.

   (b) What advantages are there in having modularity in your chosen example?

15. What other measurable quantities could be used to characterise complex networks? See, for example, Ref.[2].

5.8 References and Further Reading

1. 'Exploring Complex Networks', by S. Strogatz,
   Nature 410 (2001) 268.

2. 'The structure and function of complex networks',
   by M. Newman, SIAM Review 45 (2003) 167.
   Available from http://arxiv.org/abs/cond-mat/0303516/

3. 'Collective dynamics of small-world networks',
   by D. Watts and S. Strogatz, Nature 393 (1998) 440.

4. Linked, by A. Barabasi (Plume, 2003).

5. 'Emergence of Scaling in Random Networks',
   by A. Barabasi and R. Albert, Science 286 (1999) 509.

6. 'Biological networks: The tinkerer as an engineer',
   by U. Alon, Science 301 (2003) 1866.

Notes

Chapter 6

Societies

Humans are usually influenced in their decisions by the action of others and the environment they are in. This makes agent-based modelling of collective human behaviour a reasonable endeavour.

Indeed, the simulation of crowds is an established and important part of the design of stadiums, transport hubs and large aeroplanes. Through simulations of realistic crowds, the location of walkways and emergency exits can be optimised before construction begins.

Such simulations are one example of social models, the topic of this chapter.

6.1 Bending a Queue

Several years ago[1] this author was in a queue (line) to purchase food from a stall in a university canteen. Fig.(6.1) shows the layout, with arrows representing people facing a certain way; the arrow with the circle represents the author.

As the queue for stall B was long, it had to bend at the wall; Fig.(6.1) shows the usual way the queues bent on a typical day, while the double arrow indicates the inflow of new people joining the queues. When queue B was long, as in the figure, people heading for stall A had to cross queue B to get to the queue for stall A.

On that particular day, queue B moved forward until the situation shown in Fig.(6.2) occurred. At that point, this author, feeling somewhat annoyed at the usual state of affairs, decided on the spur

---

[1]Circa the year 2010.

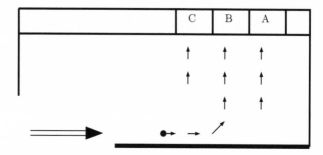

Figure 6.1: Queue bending experiment: Usual state.

of the moment to shift position slightly so that he faced in the direction shown in Fig.(6.3); hoping to re-direct the queue towards a less congested area.

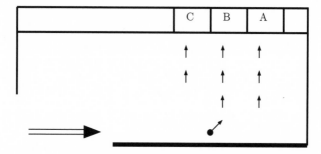

Figure 6.2: Queue bending experiment: 'Critical' state.

The next person who came to join queue B hesitated just a fraction of a second on encountering the unusual situation depicted in Fig.(6.3) before falling in behind the author — very soon the queue had bent as shown in Fig.(6.4), a very atypical situation! (Notice the inflow of people was still from the left).

When the author mentioned this real-life example to his Simplicity class, some of the students were not convinced 'queue-bending' would work if they tried it[2].

---

[2]One of the students felt that the queue-bending succeeded in my case only because I was a faculty member, while most others in the canteen then were students.

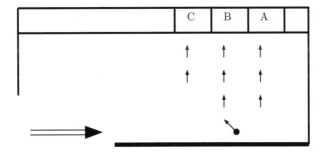

Figure 6.3: Queue bending experiment: Author changes position.

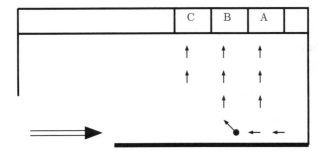

Figure 6.4: Queue bending experiment: The queue changes direction!

However, over the next few months at least two of the students independently repeated the experiment at different locations (including an airport lounge) and reported to me the hilarious results: They could bend queues with ease!

This experiment illustrates that humans do follow relatively simple rules, either implicitly or explicitly, in their daily life. In this particular situation, the rules might be to 'minimise confrontation', or to just 'go with the flow', especially if no difficulties, delays or losses are involved. Using the essence of those rules allows for the construction of relatively simple agent-based models.

6.2 Ants

One of the most studied social insects is the ant [1]. As in the case of other social insects, such as the bee or termite, there are at least two scales in the system: At the level of the individual ant, the behaviour is simple (though sometimes apparently random) while at the level of the colony one sees cooperative phenomena that are self-organised.

Without a preconceived global plan or internal supervision, ants can construct elaborate nests and forage for food, apparently using only simple local interactions among individuals [1, 2].

The model of Kirman [3] uses an element of 'randomness' to reproduce the behaviour of ants in a particular experiment. Such 'randomness' is an asset as it enables the colony to adapt to changing situations, for example in exploiting new food sources that might become available. In other words, the ant colony processes its options in *parallel*, continually exploring better solutions.

Furthermore, there is also *redundancy* in the system as many ants perform the same function and can back up their activities if the need arises. The redundancy is obviously crucial for the survivability of the colony.

There is much we humans can learn from real and simulated ant colonies, see Ref.[1] and the exercises.

6.3   Motives and Behaviour

In his book, *Micromotives and Macrobehaviour*, Thomas Schelling discussed a cellular automaton model that had two types of agents, labelled 'red' and 'blue'. The simulation started with an initial random distribution of the agents on a two-dimensional lattice [4, 5].

The movement rules were as follows: At each time step, each agent could move to a new available square, as long as the vacant square had a certain minimum number of neighbours of the same colour as the incoming agent. That rule might be motivated, for example, as follows: Migrants to a new city naturally look for familiarity.

Although the preference showed by the agents was slight — they did not explicitly seek squares where the majority of their neighbours would be of the same colour — the final result of the simulation showed an almost total segregation of colours.

Clearly, a similar conclusion could have been obtained by using a simpler set of rules: Each agent wishing to be surrounded by a majority of neighbours of the same colour.

This example illustrates how difficult in can be to guess the actual micro-motives from the observed macro-behaviour. Furthermore, an emergent macro-behaviour, such as total segregation, can become entrenched through positive feedback and eventually have consequences that were never intended.

6.4 The Sugarscape Model

In their book *Growing Artificial Societies*, Epstein and Axtell discuss a cellular automaton model of social agents that illustrates a number of patterns similar to those observed in real societies.

A noteworthy feature of the Sugarscape model is the heterogeneity of agents: Agents have different hereditary features such as metabolic rate and vision. They also have different ages and inherited wealth.

One of the numerous examples of emergence from their Sugarscape model [6] is the *Pareto Law*, a power-law probability distribution, $P(x) \sim x^{-a}$, which describes how wealth is distributed in actual societies.

The Pareto law has been found to emerge in many other mathematical models, suggesting that it is a fairly robust pattern which can arise in different circumstances. This might explain why it is also observed in various data sets.

6.5 Summary

While humans are certainly complicated beings who pride themselves on their intelligence, consciousness and free-will, most of them get through daily situations using pragmatic rules.

Allowing for heterogeneous agents, as in the Sugarscape model, and some degree of randomness, as in Kirman's model, should enable the creation of realistic social models.

6.6 Exercises

More exercises involving social models are in the next chapter.

1. What is free-will? When was the last time you exercised it?

(a) Analyse the decisions you made in the last 24 hours to see if most of them can be classified as examples of 'optimisation procedures' for saving time and money, maximising pleasure and comfort, or minimising annoyance and pain.

(b) Use the question in the last part to conduct a survey among your friends.

(c) What do your results suggest about modelling human systems?

2. Bend a queue[3]; become a believer!

(a) Suggest an agent-based model that might simulate your queue-bending experiment.

(b) Implement your model using NetLogo or other software. Observe the results and refine your model to reproduce your experimental results.

(c) Are your final rules realistic?

(d) Use your refined model to make some predictions, and then carry our further experiments[4] to test those predictions.

3. In Ref.[7] an experiment is described in which two identical food sources were placed equidistant from an ant nest. The food supplies were constantly replenished so that they remained the same as the experiment progressed. The distribution of ants between the two sources was then analysed and compared with theory. *A priori* one might have expected that the long term behaviour of the foraging ants would be largely determined by the actions of the first few, see Sect.(3.1), implying that the proportion of ants visiting the two food sources would remain stable.

However in reality the proportion of ants visiting the two sites fluctuated, showing not only expected small variations but also occasional large and rapid swings.

A. Kirman suggested a theoretical model suggested to explain the ant's behaviour [3]. An ant leaving the nest follows one of

---

[3] All experiments involving human subjects are unpredictable: Proceed at your own risk, with caution and common sense.

[4] Again, at your own risk. Proceed with caution and common sense.

three rules:
(i) It can revisit the food source it last encountered,
(ii) It can be recruited by a returning ant to visit the other source,
(iii) It can act independently and visit the other source.

  (a) Motivate the above rules, in particular rule (iii).

  (b) Are the rules compatible with the discussion in Sect.(3.1)?

  (c) Create an analogous model to explain the behaviour of a human system.

  (d) Compare your model with some examples in Refs.[3, 7].

4. Your friend Complicado argues that agent-based models of social phenomena can never be relevant to the description of real world situations because "Humans are not automatons living on a two-dimensional grid, obeying simple deterministic rules. Humans are complex beings, with free will, beliefs and emotions!" Comment critically on your friend's objection, using explicit examples for clarity.

5. Describe concisely a simple model that could simulate spontaneous human *herd behaviour* (that is, self-organisation), such as the adoption of some fad. State the rules of your model explicitly and provide motivation for them.

6. Discuss concisely an example of an agent-based model where humans have *bounded rationality*, and why the results of that model are expected to be different from one in which the agents have full information about their environment and can fully process that information.

7. Describe concisely a simple cellular automaton model that describes the spread of cultural traits in a society through the interaction of individuals. Also indicate how the model may accommodate 'leaders' who influence others but are themselves not influenced by others.

8. Implement one or more of your models from the last few questions on NetLogo or other software. Refine your models, if necessary, based on the results of the simulation.

9. In Schelling's model, an initially random distribution of agents of two colours segregates. This is opposite to what happens to the molecules of two 'colours' discussed in Sect.(4.7.1).
   Does Schelling's model violate the Second Law of Thermodynamics? Explain in detail.

10. Read Ref.[2]. An ant colony can be said to be a decentralised system consisting of (i) many, (ii) simple agents, (iii) with simple local interactions, (iv) using positive feedback, (v) and some randomness, to arrive at emergent behaviour that appears intelligent, adaptive and robust.

    (a) Explain how the achieved goals depend on the five properties highlighted above.

    (b) What do parallelism and redundancy mean with regard to an ant colony?

    (c) Discuss the advantages of parallelism and redundancy in the functioning of the human brain and compare with the ant colony.

    (d) Is the human body with its millions of cells similar to an ant colony in being a *bottom-up* emergent system, or is it a planned *top-down* system with the DNA as the control center?

11. Ant colonies.

    (a) Discuss the respective roles of self-organisation and natural selection in the development of an ant colony.

    (b) It is sometimes stated that positive feedback helps an ant colony organise while randomness enables the colony to remain adaptable. Explain clearly, with reference to an ant colony, what the terms 'positive feedback' and 'randomness' mean and how the stated goals of organisation and adaptability are achieved.

    (c) It has been suggested that inferior products can dominate a marketplace because of 'technological lock-in' caused by positive feedback. Do you think ant colonies similarly experience 'bad path lock-in' when searching for paths to their food sources? Elaborate on your answer with clear reasoning.

(d) Could there be another explanation for the seemingly occasional 'random' behaviour of some ants?

(e) How much 'randomness' is essential for adaptability of the colony? Can it be counter-productive?

(f) What lessons, if any, are there for human organisations and societies from this discussion of ant colonies?

12. (a) Your friend Critico protests: "Agent-based models in the social sciences treat human behaviour simplistically. Any resemblance of results obtained in those models to observed phenomena is either due to some fudging (artificial adjustment of rules or parameters) or pure coincidence. Thus, such models are irrelevant to our understanding of real-world social phenomena". Comment critically on your friend's opinion.

13. 'Ant colony optimisation' (ACO), first introduced by M. Dorigo [8], is an example of a biologically inspired algorithm used for solving some difficult computational problems. However, the algorithm differs in several aspects from Nature's solution, for example the virtual ants place their pheromone after their tour, the quantity deposited depending on the length of that tour.

(a) Give another example of a significant difference between virtual ants in the ACO algorithm and the way real ants solve a similar problem.

(b) Computer scientists have looked at various examples in Nature for inspiration. Describe briefly another biologically inspired algorithm (or computational method) and the type of problem relevant to humans it is used for.

(c) Discuss one advantage of the algorithm you suggested in the last part over conventional (non-biologically inspired) methods.

14. A Simplicity student has been observing an area that has a right-angled concrete pedestrian path bounding a grassy patch. She notices that after a few weeks, a clear 'short-cut' has been created by people who walk across the grassy patch instead of following the concrete path. She exclaims "The short-cut is self-organised!"

(a) Assuming that the short-cut is indeed self-organised, suggest a concise yet plausible agent-based model that could lead to that result. You should also clearly explain the motivation behind any rules you introduce.

(b) Test your model on NetLogo or other software. Refine it if necessary.

15. At certain sporting events, a propagating 'human wave' sometimes forms when adjacent groups of spectators stand and sit down successively. Is such a wave self-organised?

16. Think of a possible real-world example of self-organisation or emergence in the social sciences or human behaviour that has not been discussed in this text. Provide arguments to support your case.

17. The Pareto Law.

(a) Describe the Pareto Law mathematically.

(b) Investigate the empirical evidence that supports the law.

(c) Study one plausible mathematical model for which Pareto's Law is an emergent law.

(d) Are there exceptions to the Pareto Law?

18. In the course of his study of three countries, a student finds that in each case the distribution of wealth among individuals follows a power law, representing a great disparity between the poor and rich. He believes that this result is unnatural and a sure sign of poor social policies in those countries. Comment on the student's opinion.

19. Governments often plan and implement policy to achieve a particular outcome, but typically there are unintended consequences.

(a) Discuss one example of a government policy that had an unintended consequence.

(b) In your opinion, why do you think the unintended consequence was not anticipated?

(c) Was the unintended consequence an example of emergence?

(d) How could a study of Complexity potentially help in the formulation of better public policies?

(e) Discuss an example of a successful case to illustrate your answer to the last part.

20. Do any human organisations operate analogously to the slime mold discussed in Chap.(3)? What would be the advantages and disadvantages of such an approach?

21. Investigate some crowd simulation software on the WWW.

6.7 References and Further Reading

1. The Ants, by B. Holldobler and E. Wilson
(Belknap Press, 1990);
Ants at Work, by D. Gordon (Free Press, 1999).

2. 'Swarm Smarts', by E. Bonabeau and G. Theraulaz,
Scientific American, March 2000, 72.

3. 'Ants, rationality, and recruitment' by A. Kirman,
Quarterly Journal of Economics 108 (1993) 137.

4. Micromotives and Macrobehaviour, by T. Schelling
(Norton, 1978).

5. 'Seeing around corners', by J. Rauch, The Atlantic,
(online) April 2002.

6. Growing Artificial Societies, by J. Epstein and R. Axtell
(Brookings Institution Press, 1996).

7. Butterfly Economics, by P. Ormerod (Basic Books, 2001).

8. Ant colony optimisation, at
http://www.aco-metaheuristic.org/about.html

Notes

Chapter 7

Dynamical Systems

"Another advantage of a mathematical statement is that it is so definite that it might be definitely wrong; and if it is found to be wrong, there is a plenteous choice of amendments ready in the mathematicians' stock of formulae. Some verbal statements have not this merit; they are so vague that they could hardly be wrong, and are correspondingly useless".
— L.F. Richardson in 'Mathematics of War and Foreign Politics' [1].

In this chapter, we study models that represent the state of a system by time-dependent macroscopic variables whose evolution is determined by differential equations.

The mathematical models may be analysed qualitatively without much knowledge of calculus, and may also be investigated using free software.

7.1   Review: Rate of Change

The equation $y(x) = mx + c$ is the analytical description of a straight line in the Cartesian coordinate system. The slope (gradient) of the line is defined as the ratio of the vertical to the horizontal displacements from a point of reference. If we start at a point (x, y) and move to a point with the x-coordinate[1] $x + \Delta x$, then

---

[1]The symbol Δ represents 'change'. So Δx is 'change in x'.

$$\text{slope} \;\equiv\; \frac{y(x + \Delta x) - y(x)}{(x + \Delta x) - x} \tag{7.1}$$

$$= \frac{m\Delta x}{\Delta x} \tag{7.2}$$

$$= m. \tag{7.3}$$

That is, the slope of the line is the constant m, irrespective of the reference point.

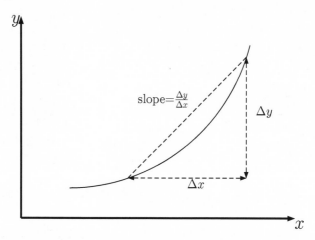

Figure 7.1: Determining the slope of a curve. The slope at a point is obtained by taking the limit $\Delta x \to 0$.

Let us now try to determine the slope of the curve $y = x^2$. This is a parabola, so its slope is not a constant. Regardless, let us proceed naively as above to see what happens:

$$\text{slope} \quad ? =? \quad \frac{y(x + \Delta x) - y(x)}{(x + \Delta x) - x} \tag{7.4}$$

$$= \frac{(x + \Delta x)^2 - x^2}{\Delta x} \tag{7.5}$$

$$= 2x + \Delta x. \tag{7.6}$$

Indeed the result is not a constant and even depends on Δx. The later dependence will disappear when we define the slope at a point as the *slope of the tangent* to the curve at that point. In that case we have to take the limit $\Delta x \to 0$, giving the result of $2x$ as the slope of the parabola at the point x. Thus, in contrast to the straight line, the slope of a curve like the parabola changes along the curve.

The following calculus notation is used to denote the result of the operation to determine the slope:

$$\frac{\mathrm{d}y}{\mathrm{d}x} \equiv \lim_{\Delta x \to 0} \frac{\Delta y}{\Delta x} \tag{7.7}$$

and is called the derivative of $y(x)$, also denoted as $y'(x)$.

We will be concerned mostly with *dynamical systems*, meaning situations where some variables of interest change with time[2]. The rate of change of the variable $x(t)$ will then be $\frac{\mathrm{d}x}{\mathrm{d}t}$, shortened with Newton's overdot notation to $\dot{x}$.

A very useful approximation when Δt is small is given by

$$\Delta x \approx \left(\frac{\mathrm{d}x}{\mathrm{d}t} \right) \times \Delta t . \tag{7.8}$$

7.2 Logistic Equation

Suppose we wished to describe how the population, x, of rabbits in a locality changes with time t. Since the number of rabbits is an integer, the mathematical expression would take the form of a *difference equation* giving x_n, the number of rabbits at time t_n, for $n = 1, 2, 3, \ldots..$

On the other hand, continuous variables are often easier to deal with; so let $x(t)$ represent the number or rabbits per unit area at time t. Treating $x(t)$ as a smooth function, we may use the methods of calculus. The simplest, and seemingly reasonable, assumption to start with would be $\frac{\mathrm{d}x}{\mathrm{d}t} \propto x$. This corresponds, from Eq.(7.8), to $\Delta x \propto x$ during short time intervals: That is, rabbits reproducing at a constant rate.

However, $\Delta x \propto x$ is an example of a *positive feedback loop*, an increase in x leading to a faster increase in x during each constant

---

[2]We will not consider here variables that change over both time and space, which would lead to 'partial differential equations'.

time interval. Thus our simple model, with a constant rate of reproduction, would lead to an exponential population growth. This is an unreasonable conclusion since the resources available to the rabbits, such as area and food, are certainly limited. Thus, one expects the rabbit population to face moderating pressures as the population increases.

So let us add a negative term, representing some negative feedback, to the right-hand-side of our original proposal, modifying the differential equation to

$$\frac{dx}{dt} = ax - bx^2, \qquad (7.9)$$

with a, b positive constants. When x is small (e.g. 0.1) the linear term dominates and the population increases. However, as x increases, the negative term becomes significant and moderates the growth. Clearly the growth stops when $ax - bx^2 = 0$, and so $x = a/b$ represents a ceiling called the *carrying capacity* of the system. (Note that the carrying capacity in this model is an *asymptotic value*, reached only as $t \to \infty$).

The choice $-bx^2$ for the moderating term is perhaps the simplest (see the 'competition' interpretation below), but it is not the only possibility, nor always the most realistic. The equation (7.9), suggested by Verhulst in the 1830's, is often referred to as the *logistic equation* and the 'S' shape of the curve describes the general trend of many processes, which involve growth and moderation, in the biological and social sciences.

As such processes tend to be quite complicated, with many factors involved, some inspired guesswork is typically required for the growth and moderation terms in the equation if one wishes for the model to fit the data reasonably well. Such models are usually called *phenomenological* as they are constructed to fit data rather than derived from first principles. Other models that represent growth followed by a maximum limit are mentioned in Ref.[2, 3].

7.2.1   A Sketch

Let us sketch the solution of Eq.(7.9) without solving that differential equation.

We will sketch x versus t, that is the *time series*. Suppose we start at $t = 0$ with some initial population x below the carrying capacity

a/b. Then using Eq.(7.8) and Eq.(7.9) we can estimate the trajectory over a short time interval $\Delta t > 0$:

$$\Delta x \;\approx\; bx \left(\frac{a}{b} - x \right) \Delta t \,. \tag{7.10}$$

Since we assumed $x < a/b$, the right-hand-side (RHS) of the above equation is positive and so $\Delta x > 0$. Clearly this trend will continue. So the trajectory $x(t)$ will be upward moving, approaching the asymptote we identified earlier at $x = a/b$, Fig.(7.2).

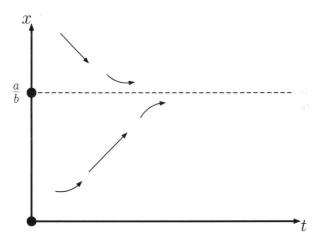

Figure 7.2: Sketch of segments of the logistic curve above and below the carrying capacity which is indicated by the dashed line.

What would have happened if initially $x > a/b$, that is, if the initial population was above the carrying capacity? Repeating the above analysis shows that the trajectory would then be a downward curve approaching the asymptote: So if the farmer placed too many rabbits (beyond the carrying capacity) in the enclosure at some initial time, the model predicts that the population will decline until equilibrium is reached at asymptotic times.

7.2.2   Fixed Points

Fixed points, or *equilibrium points* are stationary points of the dynamics, corresponding to values of x that do not change with time. To find these we set $\dot{x} = 0$.

For the logistic equation, we have two fixed points. One is at $x = 0$, this is physically sensible since if there are no rabbits, none can be generated spontaneously[3].

The other fixed point is at $x = a/b$, the carrying capacity.

However, the properties of the two fixed points are very different as we can see from the time series sketch we have done. The fixed point at $x = 0$ is *unstable* (or *repulsive*) as small perturbations create a trajectory moving away from $x = 0$.

The fixed point at $x = a/b$, on the other hand, is *stable* (or *attractive*).

Notice how we were able to deduce much about the possible outcomes using qualitative methods and elementary algebra — we did not have to solve the differential equation.

7.2.3   Competition

Since $x(t)$ represented the number of rabbits per unit area, it is therefore proportional to the probability of finding a rabbit in a unit area at time t. Thus, $x^2 = x \cdot x$ would be related to the probability of one rabbit encountering another in the same locality.

This interpretation of x^2 allows us to view the negative feedback term in the logistic equation as arising from the competition between rabbits for resources at the same location.

7.2.4   Limitations

The logistic model Eq.(7.9) contains a number of explicit and implicit assumptions that restrict its applicability.

For example, it only deals with spatially homogeneous, but time-varying, population densities represented by $x(t)$. If the actual density is not spatially homogeneous, then $x(t)$ can be taken to represent the mean spatial density[4].

One way to represent inhomogeneity, both in its spatial profile, and in the character of the agents, is to use agent-based models, which also allow for the observation of possible emergent properties in the system, see Sect.(6.4).

---

[3]Insert your joke or speculation here (*sic*).

[4]In physics, a similar strategy leads to 'mean field equations' in many applications.

However, agent-based models typically require more computational power, and so they are limited to simulations of relatively small population sizes. In contrast, differential equations modelling the mean evolution, like the logistic equation, could become increasingly better approximations for larger populations.

7.3   Modelling the Arms Race

Richardson [1] found that the defence expenditures, x and y, of the two competing European groups of nations (Franco-Russian versus Germany-Austria-Hungary) prior to World War I could be modelled well by the coupled differential equations

$$\frac{\mathrm{d}x}{\mathrm{d}t} = ay, \tag{7.11}$$

$$\frac{\mathrm{d}y}{\mathrm{d}t} = bx, \tag{7.12}$$

where a, b are positive constants. The equations imply that the rate of growth of the defence spending of each group is proportional to the actual expenditure by the competitor.

It is easy to deduce (as in our analysis of the logistic equation with $b = 0$) that Richardson's model represents an unbounded growth in the total budgets of the nations; in reality this did not happen because of the eventual outbreak of war.

Generalisations of the Richardson model are discussed in the exercises.

7.3.1   Phase Plane Portrait

It is useful to obtain a visual representation of the solutions of coupled equations such as Richardson's. Since there are two variables, we will plot $y(t)$ against $x(t)$, which yields a convenient *phase plane* portrait[5].

Each coordinate point $(x(t), y(t))$ in a phase plane plot represents the solution at a particular time t and a trajectory on that plane represents how the solution evolves with time.

---

[5]If there were more variables, we would obtain a 'phase space' plot.

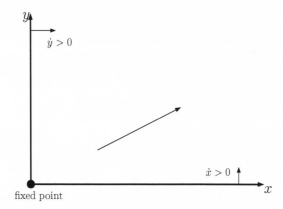

Figure 7.3: Phase plane sketch for the primary Richardson model. In this case the nullclines happen to coincide with the axes.

A phase plane plot is a powerful tool that provides useful qualitatively information about solutions to coupled differential equations, without having to explicitly solve the equations.

The strategy to sketch the trajectory is similar to that used for sketching the solution to the logistic equation: We will estimate the changes, Δx and Δy, to an initial point (x, y) in a short time interval $\Delta t > 0$. Recall from Sect.(7.2.1) and Eq.(7.8) that this requires us to know the regions where $\dfrac{dx}{dt} > 0$ and $\dfrac{dy}{dt} > 0$. The boundaries where those rates of change switch sign are called *nullclines*.

7.4   Creating a Phase Plane Plot

1. Delimit the *a priori* ranges of the variables, x and y, and any parameters (positive or negative) that appear in the equations. For example, if the variable x represents population densities, then $x \geq 0$ and if the parameter a in Richardson's equation is positive, then $a > 0$.

2. Draw appropriate axes for the variables. For example, for the Richardson model these would be the x and y axes for the first quadrant, since the physical range of the variables is positive.

3. Determine the *nullclines*, that is, the two lines corresponding to

the separate equations $\dfrac{\mathrm{d}x}{\mathrm{d}t} = 0$ and $\dfrac{\mathrm{d}y}{\mathrm{d}t} = 0$. Draw the nullclines on the graph. Mark the different regions demarcated by the lines, for example, one region might have $\dot{x} > 0$ and $\dot{y} < 0$. (Note that there might be more than one possible solution to the nullcline equations, $\dot{x} = 0$ and $\dot{y} = 0$; in that case you would need a separate phase plane plot for each distinct possibility.)

4. Pick a generic point in each region demarcated by the nullclines and determine how it would evolve a short time later. (For example, in Fig.(7.4) this is indicated by the diagonal straight arrows that are determined by their horizontal and vertical tails. The tails show the direction of Δx and Δy).

5. Sketch representative continuous trajectories that connect the different regions. (Note that two different trajectories cannot cross, as otherwise that would lead to a non-unique solution to the coupled equations emanating from the crossing point—such a scenario is not possible if our equations are non-singular.)

6. Indicate the fixed points, which correspond to the simultaneous equations $\dot{x} = \dot{y} = 0$. From your sketch you can often determine the nature of the fixed points (stable, unstable or saddle[6]).

7. From your sketch you can often determine, qualitatively, the long term behaviour of the dynamics (that is, what would happen if you waited long enough).

Note again that sometimes more than one phase plane plot is possible for the same set of coupled equations if the parameters (such as a, b, c) are not numerically fixed and if the nullclines can have qualitatively different positions relative to one another.

You can also use free software, such as Ref.[4], to obtain phase plane plots of the solutions to the coupled equations, but such software typically requires you to choose numerical values for the parameters (which means you might miss out on qualitatively different solutions if the numerical parameters you vary do not sample the different possible solution sets).

---

[6]A saddle point has features of a stable (attractive) point in some directions, while being unstable (repulsive) in other directions.

7.5   Competition between Species

Let us now study a model describing the competition between two species for the same resource. Suppose rabbits (R) and sheep (S) compete for and feed on grass in an enclosure [5]. If the land area is huge and the grass essentially unlimited, then logistic constraints are not relevant.

A simple model for their population densities is

$$\frac{dR}{dt} = aR - bRS = R(a - bS), \tag{7.13}$$

$$\frac{dS}{dt} = cS - dRS = S(c - dR), \tag{7.14}$$

where a, b, c, d are positive constants which can be given the following interpretation: The parameters a and c determine the reproduction rates of the rabbits and sheep respectively in the absence of mutual interaction between them.

With R and S being (proportional to) the probability of finding a rabbit/sheep at a certain location, then the probability of a rabbit meeting a sheep at the same location, and hence competing for the same grass, would be proportional to the product RS. Thus the negative feedback terms in the equations represent the effects of the competition.

Let us examine the dynamics through a phase plane portrait, with R along y-axis. Note that $R, S > 0$ so we should only sketch in the first (positive) quadrant. Next, determine the nullclines and demarcate the regions, for example, $\dot{R} > 0 \Rightarrow S < a/b$. The qualitative sketch of trajectories is shown in Fig.(7.4).

You should determine the fixed points by solving the two fixed point equations simultaneously in a systematic manner. The first fixed point equation implies $R = 0$ or $S = a/b$. Substitute each option separately into the second fixed point equation to get $S = 0$ and $R = c/d$ respectively.

Thus the two fixed points are at $(S, R) = (0, 0)$ (unstable) and $(S, R) = (a/b, c/d)$ (saddle). The indicated nature of the fixed points is determined from the sketch of the trajectories in their neighbourhood.

Since we have boundaries at $R = 0$ and $S = 0$, we should ask what happens if one starts on one of the boundaries; for example, starting

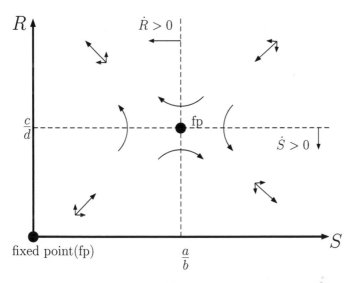

Figure 7.4: Phase plane sketch for the competition model: The dotted lines are the nullclines which demarcate the plane. The regions $\dot{R} > 0$ and $\dot{S} > 0$ are indicated. Straight arrows with horizontal and vertical tails indicate the general direction of trajectories in one of the demarcated regions. The curved arrows indicate the change in direction of trajectories as they cross a nullcline. Given this information, you can sketch representative continuous trajectories.

on the boundary $R = 0$ (the horizontal axis) equation (7.13) indicates R remains zero (a physically reasonable conclusion) while equation (7.14) indicates an ever increasing value for S, which is consistent as now the sheep have no competition, and we have ignored logistic constraints.

So what does this model predict? Firstly, rabbit and sheep cannot co-exist in stable equilibrium: Depending on the initial conditions, either one or the other will dominate the enclosure.

Of course, whether or not this model represents reality can only be answered when we have compared its conclusions with empirical data. However, even if the data were to agree with the model's predictions for some small or intermediate time scales, its prediction of an unbounded growth for one or the other species at long times is not reasonable and requires us to modify the model, as we do in the next section.

7.5.1   Including Logistic Constraints

Let us make the previous model slightly more realistic by adding a logistic constraint on the sheep population (in the exercises we will consider the more general case when both species have a logistic constraint).

$$\frac{\mathrm{d}R}{\mathrm{d}t} = R(a - bS)\,, \tag{7.15}$$

$$\frac{\mathrm{d}S}{\mathrm{d}t} = S(c - dR - eS)\,, \tag{7.16}$$

Let us examine the new phase plane sketch: The nullcline for the first equation is unchanged, but the second nullcline is now tilted since $\dot{S} > 0$ implies $R < (c - eS)/d$, see Fig.(7.5).

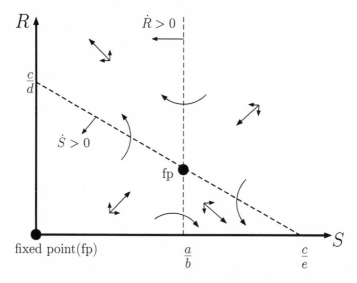

Figure 7.5: Phase plane sketch for the competition model with one logistic constraint.

How many fixed points are there? Again doing this systematically, we have from the first fixed point equation $R = 0$ or $S = a/b$ (always solve the simpler equation first). The $R = 0$ option when used in the second equation gives $S(c - eS) = 0$, thus giving two fixed points $(S, R) = (0, 0)$ (unstable) and $(c/e, 0)$ (stable). The other option from the first equation, that is, $S = a/b$ implies $R = (c - ea/b)/d$ and this corresponds to the point where the two nullclines intersect. So altogether there are three fixed points.

Notice what happens at the boundaries: The logistic constraint in the sheep equation makes that population bounded. Unfortunately, according to this model, the rabbits and sheep still cannot co-exist in a stable equilibrium.

Further generalisations are discussed in the exercises.

7.6   Modelling Epidemics

One of the simplest epidemic models to describe the spread of infections within a population was developed by Kermack and McKendrik [6]. Their *SIR model* divides the population into three classes:

- S(t) : The susceptible group of individuals, who though healthy can become infected.

- I(t) : The infected group, who can spread the disease to others.

- R(t) : The removed group consisting of dead, isolated, or recovered but now permanently immune individuals.

The coupled system of equations is

$$\frac{\mathrm{d}S}{\mathrm{d}t} = -aSI\,, \tag{7.17}$$

$$\frac{\mathrm{d}I}{\mathrm{d}t} = aSI - bI\,, \tag{7.18}$$

$$\frac{\mathrm{d}R}{\mathrm{d}t} = bI\,, \tag{7.19}$$

where a, b are positive constants denoting the infection and removal rates. The first equation implies that the rate at which individuals transfer from the susceptible to infected list is proportional to the size of the two classes. This assumes that the two classes mix uniformly.

The second term of the second equation says that infected individuals are removed with the same probability, clearly a simplifying assumption. Another assumption that has been made is that the total population is constant other than the removal due to the disease represented by the third equation (so there are no births or immigrants).

In an epidemic, the infected class of individuals grows [7]: This requires that $\dot{I} > 0$, which then implies $S > \frac{b}{a}$.

That is, the initial number of susceptible individuals must exceed a threshold value for an epidemic outbreak! This has an implication for vaccination policies, as not everyone needs to be immunised in order to prevent epidemics — all that is required is that the number of un-immunised (the susceptibles) remain below the threshold value. The actual numerical value for the threshold varies for different diseases [7].

7.7   Explosive Social Change

In his book, Epstein [7] discusses a model for revolutions (or explosive social change) which is directly analogous to the epidemic model above. Now the 'disease' is a revolutionary idea, $I(t)$ represents individuals who have become 'infected' with that revolutionary idea, and $S(t)$ is the class of people susceptible to the vision propagated by the infected. The removed class, $R(t)$, consists of infected individuals who are isolated and confined by the political establishment. The establishment tries to minimise the infection rate, a, and maximise the removal rate, b, while the revolutionaries try to maximise a and minimise b using time honoured techniques, see Ref.[7] for details.

7.8   The Predator-Prey System

Let us now study a phenomenological model describing the interaction between a species of predators and their prey as described by Volterra and Lotka [8].

The model below describes (qualitatively) the observed cyclic variation in the population of two species of fish: The predator (big-fish) population denoted by P and the small-fish population denoted by S. The equations are

$$\frac{\mathrm{d}S}{\mathrm{d}t} = S(a - bP), \tag{7.20}$$

$$\frac{\mathrm{d}P}{\mathrm{d}t} \;=\; P(cS - d)\,, \tag{7.21}$$

where a, b, c, d are positive constants that can be given the following interpretation: The parameter a represents the reproduction rate of the small-fish in the absence of predators, while b is a measure of the likelihood that a predator will encounter and eat a small-fish. Thus, the first equation(7.20) describes how the potential exponential growth of the small-fish population is moderated by a predator.

Similarly, the parameter d in the second equation describes the rate of decrease (death) of the predator population in the absence of food (small-fish), while the parameter c is a measure of the rate at which the predator population increases by meeting and feeding on the small-fish.

The coupled differential equations can be solved numerically and typically display periodic solutions [5]. On the phase plane, the trajectories cycle around the fixed point at $P = a/b$ and $S = d/c$.

That fixed point has some counter-intuitive features which highlight the intricate nature of non-linear systems. For example, suppose one wanted to increase the equilibrium population of small-fish [9]: Naively one might think of increasing the small-fish reproductive rate, a, to achieve this, but this is incorrect as we see from the solution $S = d/c$.

7.9 Summary

Similar mathematical models appear in many different contexts, so that insight obtained in one arena can sometimes be transferred to another.

The same phenomenon may be studied using differential equations for averages or agent-based models, each method having its advantages and disadvantages.

7.10 Exercises

Please note: The models in the exercises below are meant mainly to illustrate concepts, tools and approximate trends, rather than being accurate reflections of realistic situations.

1. Logistic equation.

   (a) Show that if x is initially small, then the approximate solution at early times is described by an exponential function.

   (b) Are there any real-world phenomena that are accurately described by the logistic equation?

   (c) Optional: Solve the logistic equation exactly and plot the solution. Compare with the qualitative sketch in Fig.(7.2).

2. The more general model of an arms race between two competing nations studied by Richardson is described by the following equations [1, 7]:

$$\frac{dx}{dt} = ay - bx + g_1, \qquad (7.22)$$

$$\frac{dy}{dt} = cx - dy + g_2, \qquad (7.23)$$

   where a, b, c, d, g_1, g_2 are positive constants. The equations are supposed to take into account three factors: Perceived external threats, internal economic pressures, and grievances against the other side.

   (a) Which terms in the equations correspond to those factors?

   (b) What happens if initially $x = y = 0$ but there are outstanding grievances?

   (c) What are the fixed points of the above system of equations?

   (d) Sketch and analyse the solution on the phase plane. Are the fixed points stable?

   (e) Verify your conclusions using a phase plane plotter.

3. (a) The Rapoport [10, 2] model for the arms race between two nations consists of coupled first-order differential equations for variables x, y with linear terms representing perceived external threats and internal economic factors. But the model also has additional non-linear growth terms representing the fact that in the absence of accurate information, nations are likely to accelerate their spending faster than a simple linear proportion of the actual spending by their opponents.

(a) Write coupled first-order differential equations to represent the Rapoport model.

(b) Analyse your equations and comment on the results.

4. For the rabbit-sheep system of the text,

(a) Investigate the system using a phase plane plotter. Compare with the qualitative phase plane analysis.

(b) Generalise the system of equations to include logistic constraints for both species. Analyse the equations using phase plane sketches of the qualitatively different cases. Identify those cases, if any, where there is a stable fixed point for co-existence of the two competing species, and explain the physical meaning of the constraints that lead to that case. Verify your conclusions using a phase plane plotter.

(c) What assumptions, mathematical/biological/physical, were made in modelling the system by the equations in the text? Are those assumptions realistic or reasonable? How may one generalise the model by removing those assumptions?

5. Consider the SIR model of epidemics for the special case where the removal rate $b = 0$. Then, noting that the population is constant at P,

(a) Show that $S = P - I$.

(b) Show that $\dfrac{\mathrm{d}I}{\mathrm{d}t} = aI(P - I)$.

(c) Do you recognise the last equation? Does it have a fixed point? Is the fixed point stable or unstable?

(d) Hence explain what would happen to the population if initially only a very few of the individuals became infected.

(e) How is the situation in the last part solved in reality?

(f) What assumptions, other than those mentioned in the text, have been made in the SIR model?

(g) How would one modify the SIR model to take into account individuals from the removed class re-entering the susceptible or infected classes?

6. In the SIR model, the R variable does not appear in the equations for $\dot{S}$ and $\dot{I}$. Analyse the $S - I$ coupled equations using a phase plane plot and obtain conclusions about the dynamics predicted by the SIR equations.

7. Develop an extension of the SIR model, to be called the 'SEIR' model so that there is an 'exposed' class which is different from the 'infectious' class. That is, people who have been exposed to the infection go through an incubation period before developing symptoms and becoming infectious. Could such a model be used to study the SARS outbreak of 2003?

8. Create a simple differential equation model which describes the growth of two entities as a result of a mutually beneficial interaction. Assume that in the absence of interaction both entities will have a natural decay/death rate. Predict the long-term behaviour of your model. Give an explicit example of a real-life situation described by your model.

9. In 1916, Lanchester [11, 7] described a model that may be used to study the battle between two armies x, y in engaged in face to face combat. The attrition rate for x depends not only on the size of y and the effectiveness of y's weapons but also on the size of x itself because an increasing number of combatants of one army in a battlefield implies an easier target for the enemy. The non-linear Lanchester equations, including further the possibility of reinforcements are

$$\dot{x} = -k_2 xy + x(r_1 - a_1 x), \qquad (7.24)$$
$$\dot{y} = -k_1 xy + y(r_2 - a_2 y), \qquad (7.25)$$

(a) Interpret each of the terms in the above equations.

(b) Study the above system of equations and comment on the results.

(c) Are the Lanchester equations a realistic description of modern battlefields?

(d) Which other system might be described by equations similar to the above?

10. Lotka-Volterra equations.

(a) Attempt a phase plane plot of the equations. What ambiguities do you face?

(b) Explore the equations using software.

(c) Modify the original predator-prey equations to include logistic constraints for the prey. Investigate the modified system.

11. A biodiversity student tentatively models the time evolution of the mass density, $x(t)$, of algae in a large lake, using the equation

$$\dot{x} = x(z - x) \tag{7.26}$$

with $z > 0$ a constant.

(a) Explain the significance of the constant z and the long-time prediction of this equation.

(b) On further study, the student finds that the long-time prediction of equation (7.26) does not fit the empirical data, and so she decides to make z a dynamical variable, $z(t) > 0$, evolving as

$$\dot{z} = z(Cx - z) \tag{7.27}$$

with $C > 0$ a constant. That is, equations (7.26 - 7.27) are now treated as coupled equations for the time-dependent variables $x(t), z(t)$.

   i. Provide plausible motivation for the terms on the right-hand-side of equation (7.27).

   ii. By sketching relevant phase space plot(s) for the coupled equations (7.26 - 7.27), use clear reasoning to deduce the condition on the constant C that would allow an ever increasing amount of algae to be supported in the long term. (You may assume that the initial conditions are such that x, z are both non-zero.)

12. A political science student models the competition between two political parties, x and y, for voter support in a large city. His initial dynamical model, with x and y representing the number (per unit area) of supporters of each party, is:

$$\begin{aligned} \dot{x} &= x(1 - 2y), \\ \dot{y} &= y(3 - 4x), \end{aligned}$$

where t is time. However, he finds that this model does not allow for a stable equilibrium with non-zero values for both of the variables, as observed empirically.

(a) Modify the above equations by adding two new terms, one to each equation, so that a stable co-existence is produced. Label as a and b the two new positive constants you introduce and deduce constraints on their values by sketching and analysing the relevant phase space plot(s) corresponding to the above equations.

(b) If party y is found to have a larger support than party x at the stable equilibrium point in part (a), what further constraints can one place on the constants a and b?

(c) Another social science student is planning on studying the same phenomenon (competition for votes) using instead an agent-based simulation. Discuss one major advantage and one major disadvantage of such an approach compared to the above method of using differential equations.

13. In the year 2049, recruitment of new students (the 'suscepti-bles') for Rambutan College takes place through their interaction with the existing students (those 'infected' with the Rambutan spirit). For her group project, a Simplicity student models the population of different groups of students by the following set of equations,

$$\dot{S} = S - 2SI, \qquad (7.28)$$
$$\dot{I} = aSI - I^2, \qquad (7.29)$$
$$\dot{Q} = bI^2, \qquad (7.30)$$

where $S(t)$ represents the group of susceptibles, $I(t)$ the infected, and $Q(t)$ the group of students who quit Rambutan College to join Duku College. The time parameter is t.

(a) What values should be assigned to the positive constants a, b? Explain.

(b) Using the values from part (a), draw a complete phase space plot for the equations (7.28,7.29) (with S along the y-axis) which incorporates the following additional information: The fixed point at the intersection of nullclines is stable.

(c) Hence deduce the behaviour of the variables S, I, Q in the long term if none of them is initially zero.

(d) Is the total population $P = S + I + Q$ constant? Explain.

14. Two independent cable TV channels compete for a share of the audience in a city. A business analyst uses a mathematical model to study the various possible outcomes of that competition. The dynamical model, with x and y representing the number (per unit area) of subscribers of each company, is

$$\dot{x} = x(a - 3x - y),$$
$$\dot{y} = y(2 - bx - y),$$

where a, b are positive constants and t the time.

(a) If in reality both companies are found to reach a stable equilibrium with each having a non-zero share of the market, deduce the possible range of values for a and b by sketching and analysing the relevant phase space plot(s) corresponding to the above equations.

(b) If company y is found to have a larger market share than company x at the stable equilibrium point in part (a), what can one say about the relationship of the constant a to constant b?

(c) Suppose the competition term in the above equations were eliminated. Then, how much market share would company y hold, in the long term, in the absence of company x?

15. A mathematical biologist models the population density of a parasite and of a certain species it infects. The equations representing the parasite and host populations are

$$\dot{w} = w(2 - 8z), \tag{7.31}$$
$$\dot{z} = z(5 + 3w - z), \tag{7.32}$$

where w, z are the variables and t the time.

(a) Which variable represents the parasite population? Explain.

(b) If initially there is a non-zero parasite density, what does the model predict for the long term? Justify your answer.

(c) Discuss one prediction (or assumption) of the model that is unrealistic and how you would modify the equations to make them more realistic. (You do not need to solve the modified equations).

(d) Find all fixed points of your model that describe only one surviving species. Identify the surviving species and its population density.

16. (a) An engineer finds that the state of a system can be described by two non-negative time-dependent variables $x(t)$ and $y(t)$ but she does not know the coupled firt-order differential equations that determine the evolution of those variables. However after much experimentation she guesses that the $x - y$ phase plane of the system has three fixed points: An unstable fixed point at $(0,0)$, a stable fixed point at $(a,0)$ and a saddle point at (b,c) with $a > b > 0$.

   (a) Sketch a plausible phase plot for the system, showing some typical trajectories.

   (b) Deduce the long-term behaviour of the system if initially it is not exactly at any of the fixed points.

17. The stability of fixed points can be studied analytically using *linear stability analysis*. Learn that technique and use it to verify the stability of fixed points you investigated previously using phase plane plots.

18. Choose a model from this chapter and recreate it as an agent-based model that you can implement using NetLogo or other software. Compare the different insights obtained.

7.11   References and Further Reading

1. 'Mathematics of war and foreign politics', by L. Richardson, Arms and Insecurity (Boxwood Press, 1960).

2. Nonlinear Physics with Maple for Scientists and Engineers, by R. Enns and G. McGuire (Birkhauser, 2000).

3. Real World Mathematics, by W.K. Ng and R. Parwani (Simplicity Research Institute, 2014).

4. A Phase Plane Plotter at
   http://www.math.missouri.edu/ bartonae/pplane.html

5. Nonlinear Dynamics and Chaos, by S. Strogatz (Westview Press, 2000).

6. 'A Contribution to the Mathematical Theory of Epidemics', by W. Kermack and A. McKendrick, Proceedings of the Royal Society A 115 (772) (1927) 700.

7. Nonlinear Dynamics, Mathematical Biology, and Social Science, by J. Epstein (Addison Wesley, 1997).

8. Elements of Physical Biology, by A. Lotka (Williams and Wilkins, 1925);
   'Variations and fluctuations of the number of individuals in animal species living together' by V. Volterra, Animal Ecology (McGrawHill, 1931).

9. The Computational Beauty of Nature, by G. Flake (MIT Press, 1998).

10. Fights, Games, and Debates, by A. Rapoport (University of Michigan Press, 1960).

11. 'Mathematics in Warfare' by F. Lanchester, in The World of Mathematics, Vol. 4, 2138 (Simon and Schuster, 1956).

Notes

Chapter 8

Fractals: Geometrical Complexity

How does one describe the outline of a coastline or the spatial structure of a snowflake? Benoit Mandelbrot [1] introduced the word *fractal* to describe such shapes that appear self-similar after a change of observation scale (magnification).

The technical term that describes self-similarity of shapes under change of observation scale is *scale-invariance*. Systems that are scale-invariant do not have any characteristic length, that is, a typical or mean length.

For example, if one observes an aerial photograph of a river with its tributaries, such as the Amazon river system, it is difficult to guess the actual size of the features unless some man-made objects are also visible. This is because the Amazon river system is approximately scale-invariant, while man-made products have a natural characteristic length: A car is about 3m long and the linear dimension of a house is about 10m. Objects that have a characteristic length scale look different at different magnifications.

8.1 Exact Fractals

We will first study mathematical fractals that are exactly self-similar at all scales. Such fractals are generated by iteration, that is, by repeating a procedure a number of times. The number of iterations is kept track of by an integer k.

8.1.1   The Koch Curve and Snowflake

This fractal is generated by iteration as follows [2]: The *initiator*, the initial or $k = 0$ step, is a unit line element. The first step, $k = 1$, called the *generator*, involves removing the middle one-third of the unit line and replacing it with two line segments each one-third in length . The generator, shown in Fig.(8.1), contains four equal line segments.

In the next step, $k = 2$, each of the four line segments of the $k = 1$ figure is replaced by the (scaled) generator, leading to a figure with 16 segments. The procedure is repeated endlessly, $k \to \infty$, to generate the *Koch curve*, Fig.(8.1).

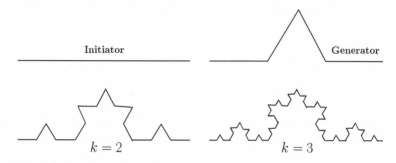

Figure 8.1: Generation of Koch Curve.

Note that the Koch fractal, with its exact self-similarity at all scales, is obtained only after an infinite number of iterations, at the $k = \infty$ step. After a large, but finite, number of iterations, the figures already look like the final fractal because of the limited resolution of our eyes.

Three Koch curves can be fitted together to form a two-dimensional figure called the 'Koch snowflake', Fig.(8.2). Alternatively, one can start with an equilateral triangle and apply the generator of the Koch curve repeatedly to each line segment.

How long is the boundary of a Koch snowflake? At each step in its generation, the length increases by a factor of 4/3 as a line segment of one-third unit is replaced by two of equal length. Therefore as $k \to \infty$, a Koch snowflake is produced with a boundary of infinite length, but with an area that is finite (bounded by the circle that circumscribes the original triangle). This unusual situation is possible

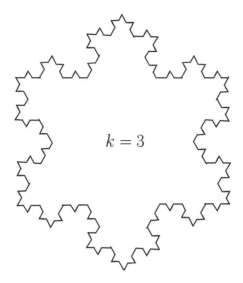

Figure 8.2: Generation of Koch Snowflake.

because the boundary of the Koch snowflake is infinitely 'kinky' (the curve consists entirely of corners).

The Koch snowflake, with its infinite length but finite area, is an unusual object from the viewpoint of Euclidean geometry where objects in finite space have finite lengths, areas and volumes. In order to characterize objects such as the Koch fractal we have to generalize our notion of dimension.

8.2 Dimensions

We live in a three-dimensional world: That is, we need three coordinates to specify the location of any point. The coordinate system we use locally might be the orthogonal (right-angular) Cartesian grid with x, y, z axes, or more practically for a global description, a location of a point in terms of latitude, longitude and altitude. When we restrict our attention to a subset of the world, we can often make do with a description in terms of fewer coordinates. For example, a point on the surface of a table can be described in terms of two coordinates.

Sometimes a description in terms of fewer coordinates is a useful and economical approximation. Consider a long thin string. It exists in three-dimensional space, but any point on it can be located by tracing out a single coordinate along the string.

In the example above we have used two definitions of dimensions. Firstly there is the *Euclidean dimension* (D_e): The number of coordinates required to specify an object in space. Secondly there is the *topological dimension* (D_t), which, roughly speaking, is a measure of the intrinsic dimension of the object. For example, a thin string has topological dimension one and an Euclidean dimension of three.

Topology is often called 'rubber geometry' because it deals only with the qualitative shape of an object. If the object is imagined to be made of rubber then by stretching (but without tearing) it can be deformed into another topologically equivalent object. Thus, a curve of any shape is topologically equivalent to a straight line, and has a topological dimension of one.

The Euclidean and topological dimensions are always integral. For characterizing fractals it is useful to introduce another definition called the *similarity dimension* [1]. To motivate this definition, consider first a unit Euclidean line, square and cube, each divided into N equal self-similar parts of linear dimension s [2].

For the line, since $Ns = 1$, each smaller part has a length $s = 1/N$. For the square, $Ns^2 = 1$, so each smaller square has length $s = 1/N^{1/2}$, while since $Ns^3 = 1$ for the cube, each smaller cube has length $s = 1/N^{1/3}$.

Definition of Similarity Dimension: If an object of unit size contains N self-similar copies of itself of linear size s, then its similarity dimension D_s is determined by the equation

$$Ns^{D_s} = 1 \, . \tag{8.1}$$

For the Euclidean objects mentioned above, $D_s = 1$ for the line, $D_s = 2$ for the square and $D_s = 3$ for the cube. These numbers are identical to the topological and Euclidean Dimensions for those objects. Let us rewrite the above equation in the form

$$D_s = \frac{\log(N)}{\log(1/s)} \, . \tag{8.2}$$

Now we can find the similarity dimension of the Koch curve. At each observation scale, the curve contains 4 self-similar copies of itself of size $s = 1/3$, so

$$D_s = \frac{\log(4)}{\log(3)} = 1.2618... \qquad (8.3)$$

Thus the similarity dimension of a Koch curve is larger than its topological dimension which is one, but smaller than its Euclidean dimension of two. Since D_s for a Koch curve is larger than that for a line but smaller than that for area, one can roughly say that the Koch curve is more than a line but not quite a plane. We are now in a position to appreciate the following formal definition of a fractal [1].

Definition of Fractal[1]: A fractal is an object whose similarity dimension is larger than its topological dimension.

Here is another example of a fractal: Mandelbrot discovered that noise in telephone lines is clustered and can be modelled as a Cantor set. The *Cantor set* is generated as follows: The initiator is a unit line element. The generator involves removing the middle one-third of the unit line. After this first step the figure consists of two line segments, each one-third in length. The procedure is repeated endlessly, each time removing the middle one-third of the remaining line segments.

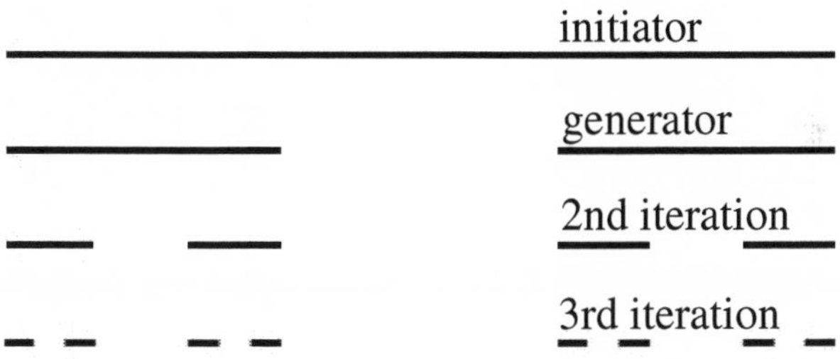

Figure 8.3: Generation of Cantor Set.

The Cantor set has an infinite number of points but it is of width zero[2]. The Cantor set is in some sense the opposite of the Koch curve: In the generation of the Cantor set, at each step the line segment was

---

[1]See the references for more sophisticated definitions.

[2]A set of points, even if they are countably infinite, has topological dimension zero; this is because the integers form a subset of measure zero on the real line [4].

made shorter by one-third while in the Koch curve it was made longer by one-third. Therefore, one would expect that the Cantor set is in fact 'less than a line', just as the Koch curve was considered 'more than a line'; see the exercises.

8.3   Natural Fractals

Natural fractals are self-similar only over a limited range, and the self-similarity is often statistical rather than exact.

An algorithm to generate natural looking fractals is obtained by modifying the iteration process of the last section to include a probabilistic component [1, 2]. Consider the generation of a random Koch curve: The initiator and generator are as before, but in the following steps ($k = 2$ onwards), the figures are obtained by replacing each line segment with the generator in such a way that the apex of the generator triangle points randomly (for example, determined by a coin toss) to either side of the original line, see Fig.(8.4). The final figure, at $k = \infty$, is closer to the shape of natural objects such as coastlines.

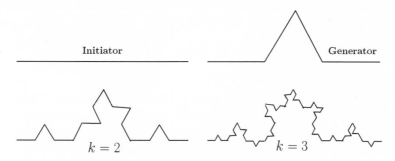

Figure 8.4: Generation of a random fractal.

Just as for exact fractals, one can introduce a dimension to characterize random fractals. One example is the *box-counting dimension*, see Fig.(8.5). In this method, the figure is covered by equal sized cells (cubes or squares) of linear dimension r. One then counts the number of cells, $N(r)$, that are needed to cover the given figure. If

$$N(r) \propto r^{-D_b} \tag{8.4}$$

as the length r is changed, one says that the distribution of points is D_b-dimensional[3]. This definition agrees with the Euclidean dimension for straight lines and planes but gives fractional values for more complicated shapes such as coastlines. Note that the equation above is of the same form as that which comes from the definition of the self-similarity dimension mentioned above.

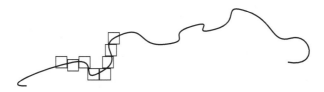

Figure 8.5: The box-counting method. Only a few of the boxes are displayed on the left.

8.4 Dynamical generation of natural fractals

Some natural fractals, such as the clusters describing a bacterial colony, can be generated by a physically motivated model called *diffusion-limited aggregation* (DLA) [5]. Consider for simplicity the formation of such a cluster in the plane, with the initial (seed) particle located at the origin $(x = y = 0)$.

Other particles are then released far from the origin, at random locations, and allowed to diffuse: Mathematically this is done using an algorithm such as a *random walk*[4] to simulate the diffusion process.

When the diffusing particle encounters the seed particle it is made to stick to it. The process is repeated with other diffusing particles, leading to the formation of a cluster. As the cluster forms, there is a greater probability for particles to stick to the ends than to penetrate the interior. This leads to the formation of a branch-like structure emanating from the origin.

Some other examples of DLA are in the growth of crystals (e.g. snowflakes) and coral reefs [3].

Another dynamical 'explanation' for the ubiquitous occurrence of fractals in nature is the idea of self-organised criticality, see Sect.(3.7).

---

[3]In general, the limit $r \to 0$ must be taken.
[4]See Sect.(4.1).

8.5   Usefulness

Look at pictures of the branching of a tree or river, the branching of air passages in the lung, the branching of blood vessels in the human body, or the folds on the surface of the brain.

Why should such fractal-like structures exist in Nature?

Recall from the example of the Koch snowflake that one can accommodate long lengths in small areas. Similarly, higher dimensional analogues accommodate large surface areas in small volumes (see exercises). This suggests that Nature chooses fractal structures to optimize functional efficiency given limited resources.

8.6   Summary

Fractals are shapes that are self-similar on multiple scales (spatial or temporal), and in general have fractional dimensions. Natural fractals are self-similar only over a limited range, and the similarity is often statistical rather than exact.

The complex geometry of fractals can be reproduced by simple algorithms iterated over many steps. However, it is rarely possible to guess what the final structure will look like until numerous steps have been performed.

Fractals have found applications in many domains, such as in data compression.

8.7   Exercises

1. Many films use miniature models for their special effects. Discuss how the human brain is tricked into thinking that those models are life sized and how this relates to the concept of characteristic length or self-similarity.

2. Natural Fractals.

   (a) Explain why fractals in Nature are self-similar only over a limited range. Is there an upper limit or lower limit or both?

   (b) Search for two examples in your environment that are almost exact fractals, and two examples of random fractals.

3. Dimensions.

   (a) Give examples of objects that are self-similar over a wide range of scale but which are not fractals.

   (b) Give an example of an object that is a fractal but does not have fractional dimensions.

4. Do you think the fractal (self-similarity) dimension of a two-dimensional projection of a tree is more than, or less than, 2? Why?

5. Koch Curve and Snowflake.

   (a) Can you draw a unique tangent at any point of a Koch curve? Why not? What is the mathematical terminology to describe such curves?

   (b) Suppose the initiator of the Koch curve was 1cm long. Approximately how long would the curve be after 100 iterations? If the curve was a piece of thread that could be stretched to its full length, could you use it to tie the Earth to its Moon?

   (c) What are the topological and Euclidean dimensions of the Koch curve?

   (d) Calculate the self-similarity dimensions of the Koch curve using the scales $s = 1/9$ and $s = 1/27$.

   (e) What is the area of a Koch snowflake if the initiator was an equilateral triangle of unit length on each side?

6. Show that the length of the initiator that remains once a Cantor set is formed is zero. Convince yourself that the Cantor set is a fractal, by showing that it is self-similar and also by computing its similarity dimensions and comparing that with its Euclidean and topological dimensions.

7. The *Menger Sponge* is a mathematical fractal in three Euclidean dimensions. It has zero volume but an infinite surface area. Some properties of the human brain are roughly similar to the Menger Sponge. Explain what those properties are and why they are biologically useful.

8. *Sierpinski carpet and Menger sponge.*

    (a) Determine the area and perimeter of the Sierpinski carpet.

    (b) Determine the topological and self-similar dimensions of the Sierpinski carpet and the Menger Sponge.

    (c) In what way are the Cantor set, Sierpinski carpet and Menger sponge related?

9. Measure the fractal dimension of a long coastline of your choosing.

10. Historically, Richardson observed that the length of some borders between countries seemed to increase when the length of the measuring instrument was reduced [3]. Can you explain in physical terms what is happening in this case?

11. What are multi-fractals? Do they occur naturally?

8.8 References and Further Reading

1. The Fractal Geometry of Nature, by B. Mandelbrot (W.H. Freeman, 1983).

2. Fractals and Chaos, by P. Addison (IOP Publishing, 1997).

3. The Computational Beauty of Nature, by G. Flake (MIT Press, 1998).

4. Nonlinear Dynamics and Chaos, by S. Strogatz (Westview Press, 2000).

5. Diffusion-limited aggregation by T. Witten Jr. and L. Sander, Phys. Rev. Lett. 47 (1981) 1400.

Chapter 9

Chaos

When the meteorologist Edward Lorenz [1, 2] was studying a simplified model of the weather, he discovered that tiny differences in the initial conditions rapidly led to very different results.

This came as a surprise as it was always assumed that small errors, such as the uncertainties inherent in real-life measurements of finite precision, would lead to small corrections.

Lorenz referred to the sensitive dependence on initial conditions as the *butterfly effect*: For example, a butterfly flapping its wings in Singapore might completely change the weather in New York!

9.1 Occurrence

Given *deterministic*[1] equations describing a system, we may informally define *chaos* as: Sensitivity of the solutions to initial conditions, causing previously nearby trajectories to rapidly diverge from each other[2].

Chaos implies that even systems with a few variables can show complicated behaviour that is practically unpredictable on long time scales.

A necessary condition for chaos is that the equations for the system are non-linear as errors in linear systems remain small if they were initially small. As most of the dynamical systems in real life are

---

[1]Precisely defined dynamics, with no randomness.

[2]For a more precise definition, which includes some reference to the attractor, see Ref.[3].

described by non-linear equations, it is expected that chaos will be commonplace.

However, even for non-linear systems, chaos might only occur for some values of the parameters that appear in the equations, and so real life non-linear systems might not always be chaotic. In other words, non-linearity is not a sufficient condition for chaos[3].

It must be emphasized that chaos is very different from randomness: The former arises in perfectly deterministic systems while the later is intrinsically non-deterministic, and the distinction between the two at the practical level can be seen by looking at the 'phase space' of the system[4], see later.

9.2   The Double Pendulum

A simple pendulum consists of a small weight supported by a very light string. The string is attached to a pivot about which it can swing. If the amplitude of oscillations is small, the equation of motion is approximately linear and the period of oscillation is essentially independent of the amplitude. The motion is described as 'simple harmonic motion' [4].

For larger amplitudes of oscillation, the equation of motion is non-linear but the oscillations are periodic and very predictable.

Consider now a double pendulum, which consists of two simple pendulums in tandem: One attaches a single pendulum to the end of another, Fig.(9.1). The equations of motion are again non-linear for large oscillations, but now the motion becomes quite irregular and very sensitive to the initial conditions. This kind of behaviour is the hallmark of chaos. See the simulation of the double pendulum on the WWW.

9.3   The Rossler System

A model displaying chaos, which is simpler than the weather model of Lorenz, was proposed by Rossler [5, 2]. The system consists of three

---

[3]See also Exercise (6).

[4]The word 'chaos' is one example where the technical definition is very different from its popular colloquial usage.

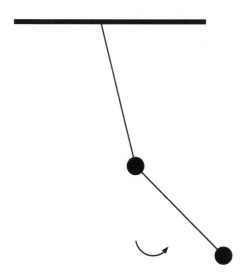

Figure 9.1: The double pendulum is chaotic for large angles of oscillation.

coupled first-order differential equations:

$$\dot{x} = -(z + y), \tag{9.1}$$

$$\dot{y} = x + ay, \tag{9.2}$$

$$\dot{z} = b + xz - cz. \tag{9.3}$$

Rossler's system has three time-dependent variables[5] represented by x, y and z. Notice the non-linear term xz in the last equation. There are also three parameters a, b and c. By fixing two of the parameters and varying the third one can study the approach to chaos. Let us fix $a = b = 0.2$ and treat c as the control parameter. Instead of plotting the time series for the system, it is useful to consider phase portraits, that is by plotting one variable against another.

Fig.(9.2) and Fig.(9.3) show the $x - y$ phase portraits (of the post-transient stage) for two values of c. One clearly sees period two and period four trajectories; the period here refers to the number of times the cycle goes around before closing. These period doubling sequences continue as c increases, leading eventually to the aperiodic chaotic state shown in Fig.(9.4).

---

[5]Recall that $\dot{x}$ is the time derivative of x.

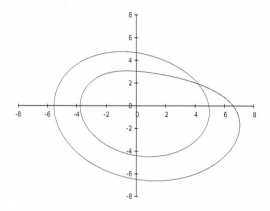

Figure 9.2: A period two orbit of the Rossler system in the $x - y$ plane.

The behaviour of the z variable in the chaotic state is illuminated by looking at the time series shown in Fig.(9.5): While the z coordinate is small most of the time, meaning that the trajectories mostly are close to the $x - y$ plane, occasionally there are large spikes in the z coordinate [2].

The full three-dimensional phase portrait for the chaotic state is shown in Fig.(9.6). This is called the *strange attractor* for the system [3]. It is an 'attractor' in the sense that the trajectories lie in a bounded region of phase space.

Though complicated, the attractor is far from random. Indeed 'strange attractors' have a fractal structure [3], that is, self-similarity at different scales! This is a clear indication that deterministic chaos is different from noise or randomness.

We noted that in the chaotic state, nearby trajectories diverge exponentially, and yet the phase portrait is bounded. How is this possible? This is done by a process of stretching of nearby trajectories on short time scales, followed by a process of folding at longer scales [5, 3]: If you follow some trajectory in the $z = 0$ plane, it will soon be lifted up at a spike of the z-component, see Fig.(9.5), and is then folded back into the $z = 0$ plane of the attractor at another point, to continue its evolution. The stretching and folding processes mix nearby trajectories in the attractor.

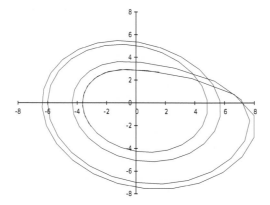

Figure 9.3: A period four orbit of the Rossler system in the $x-y$ plane. The actual curve is smooth. It appears jagged here only because the interpolating points used to generate the figure were far apart.

9.4 Summary

If a dynamical system becomes chaotic at some parameters of physical interest, then that is usually an unwelcome occurrence. The extreme sensitivity implies a lack of practical predictability of the long-term behaviour of the system even though the equations of motion are completely deterministic.

Nevertheless, long-term unpredictability due to chaos does not preclude short-term predictability and usefulness. Furthermore, one might be able to exploit the sensitive dependence of chaotic systems to initial conditions by controlling those systems: Very small perturbations to the system can be used to completely change its long-term behaviour.

Chaos arising in deterministic systems is quite different from noise in random (stochastic) systems. This is because one still sees patterns in the phase space plots for chaotic systems which are absent for random systems. A technique for distinguishing whether a given time-series has a deterministic underlying basis is through the method of attractor reconstruction [3].

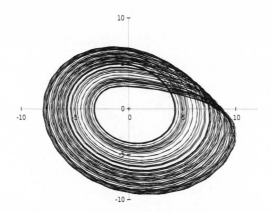

Figure 9.4: The $x - y$ plane slice of the Rossler phase portrait in the chaotic state.

9.5   Exercises

1. Use your favourite software to explore the Rossler system (time series and phase plots) for different values of the control parameter c. Note that by plotting only later points of the iteration, the transient part of the orbit is neglected. Plot also the transient parts of the orbits to see the approach to the attractors.

2. Look up the equations for the Lorenz system and study it using software.

3. The *Lyapunov exponent* and *Lyapunov time* are used to characterise chaotic systems. Investigate.

4. It is often stated that our Solar System is chaotic. What would that imply? Should we worry?

5. There have been suggestions [6] that chaos may be used to encode messages which need to be sent privately. Read about and discuss this possibility.

6. The Poincare-Bendixon theorem implies that strange attractors cannot arise in continuous two-dimensional dynamical systems, even if they are nonlinear. This means that various models of Chap.(7), such as Richardson's two nation arms race model and

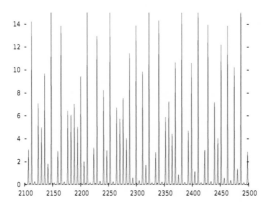

Figure 9.5: Time series for the z variable of the Rossler system in the chaotic state.

the basic Lotka-Volterra model, cannot become chaotic. Investigate which models from Chap.(7) could potentially become chaotic.

7. Newton's Second Law leads to equations of the form $\ddot{x} = f$, where the double dot refers to the second time derivative. By introducing a new variable through $\dot{x} \equiv y$, the original second-order differential equation can be written as two coupled first-order equations.

   (a) Look up the equations for double pendulum and re-write them as first-order equations.

   (b) Would you expect the system to be chaotic? Why?

   (c) Investigate the system using software.

9.6 References and Further Reading

1. 'Deterministic nonperiodic flow' by E. Lorenz, Journal of the Atmospheric Sciences 20 (1963) 130.

2. Nonlinear Physics with Maple, by R. Enns and G. McGuire (Birkhauser, 2000).

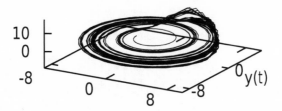

Figure 9.6: The three-dimensional strange attractor of the Rossler system.

3. Nonlinear Dynamics and Chaos, by S. Strogatz (Westview Press, 2000).

4. Real World Mathematics, by W.K. Ng and R. Parwani (Simplicity Research Institute, 2014).

5. 'An equation for continuous chaos', by O. Rossler, Physics Letters A 57 (1976) 397.

6. 'Circuit implementation of synchronized chaos with applications to communications' by K. Cuomo and A. Oppenheim, Phys. Rev. Lett. 71 (1993) 65.

Chapter 10

Quantifying Complexity

This chapter provides a heuristic introduction to some measures of complexity. For details and other common measures please see the references.

10.1 Computational Complexity

Computer scientists are naturally keen to find efficient problem-solving algorithms.

The *computational complexity* of a problem of size N is the amount of time t required to solve it. If t is bounded by a polynomial in N, that is $t \sim N^a$ where a is a constant, then the problem is said to be in the *complexity class P*.

Note that we are talking about the best (time-saving) deterministic algorithms for solving the problems.

An example from the P class would be to find the product of two N digit numbers: The time required is bounded by a polynomial.

By contrast, a problem from the EXP class would need $t \sim 2^N$, that is the amount of time required would grow exponentially with the size of the problem. An example from this class would be the evaluation of a position in a game of generalised chess[1], played on a $N \times N$ board.

The problems that most interest researchers are those in the *non-deterministic polynomial* (NP) class. For such problems, it might be

---

[1] Usual chess is played on a 8×8 board.

possible to first guess a solution that can then be checked in polynomial time; an example is the factorisation of a large number.

It is unknown if NP problems can be solved deterministically in polynomial time. One of many famous research problems, for which you could win a million (US) dollars in prize money, is to determine if the class NP is the same as P [3].

10.2   Algorithmic Complexity

The *algorithmic complexity*[2] of a data set is the length of the shortest algorithm (precise instructions) that can generate that data.

For example, consider the list of natural numbers from one to one billion. That data set can be generated by a short algorithm of the form: 'Start. $N = 0$. Increase N by 1. If $N > 10^9$ stop; if not, print N. Continue loop.'

Both the data set and the computer algorithm are naturally coded in binary form (a string of digits using the integers 0 and 1). In the above example, the algorithm is obvious shorter than the data set it generates.

A data set that has low algorithmic complexity is highly compressible, either because of a pattern or redundancy, or both. By contrast, a data set that is mostly random would have a very high algorithmic complexity because it would be difficult to summarise the data other than listing it directly.

Thus, a data set might have high algorithmic complexity but be uninteresting.

The measure in the next section is one attempt to address the difficulty of characterising 'complexity'.

10.3   Logical Depth

In the study of complex systems, we saw that simple rules could give rise to interesting emergent properties that we could not have easily guessed from the rules themselves.

Bennett [4] introduced the concept of *logical depth* to quantify the amount of effort it would require to go from the rules (the short algorithm) to the final result.

---

[2]Also known as Kolmogorov-Chaitin complexity [2].

A data set with low logical depth would follow quickly from the rules while one with high depth would require much effort even if the rules were simple.

This measure seems to capture some essence of 'complex systems' as we have introduced them.

10.4 Exercises

1. Find other examples from the class P, NP and EXP. What are *NP-complete* problems?

2. It is sometimes stated that the ACO algorithm, mentioned in the exercises of Chap.(6), can solve NP problems in polynomial time. How is this possible?

3. The laws of physics, such as Newton's law of universal gravitation, attempt to summarise the underlying principles behind diverse phenomena. Relate physicists' attempt to find a few simple universal laws to the concept of 'algorithmic complexity of Nature'.

4. Could algorithmic complexity, or the other measures in this chapter, be used to distinguish between the notions of 'complex' and 'complicated'?

5. Do fractals have low or high logical depth? Explain.

6. Find examples of systems with low algorithmic complexity and low logical depth.

7. Information and Uncertainty.

   (a) 'Entropy', is used in physics as a measure of disorder (or uncertainty) and in computer science as a measure of information. How is uncertainty related to information?

   (b) Is the entropy measure related to any of the other complexity measures in this chapter? See Ref.[5].

8. Examine other suggested measures of 'complexity', and discuss their advantages and short-comings.

10.5   References and Further Reading

1. The Computational Beauty of Nature, by G. Flake
   (MIT Press, 1998).

2. 'On Tables of Random Numbers' by A. Kolmogorov,
   Sankhya Ser. A. 25 (1963) 369;
   'On the Length of Programs for Computing Finite Binary
   Sequences', by G. Chaitin, Journal of the ACM 13 (1966) 547.

3. www.claymath.org/millenium-problems/p-vs-np-problem

4. 'Logical Depth and Physical Complexity', by C. Bennett,
   in The Universal Turing Machine
   (Oxford University Press, 1988).

5. 'Entropy, information, and computation', by J.Machta,
   American Journal of Physics 67, (1999) 1074.

Chapter 11

Exploring Complexity

Here are some pointers and suggestions for readers who wish to go beyond this text[1].

11.1 Learning Complexity

A starter list of books, articles and websites is provided below. Most of the books have an inside preview at http://books.google.com/ or www.amazon.com/, and related books may be found through the latter website.

Other relevant weblinks and references will be posted on this book's webpage at **www.simplicitysg.net/books**.

11.1.1 Introductory

1. Wikipedia (www.wikipedia.org/) is a good place to start for a quick overview and links to other references.

2. Emergence, by S. Johnson (Simon and Schuster, 2012).

3. Linked, by A. Barabasi (Plume, 2003).

4. Sync, by S. Strogatz (Hyperion, 2012).

5. 'Swarm Smarts', by E. Bonabeau and G. Theraulaz, Scientific American, March 2000, 73.

---

[1]Keep in mind the need to cross-check the accuracy of crucial information.

6. 'Seeing around corners', by J. Rauch,
   The Atlantic (online), April 2002.

7. The Complexity Explorer: Free on-line courses at
   http://www.complexityexplorer.org/

8. The Sciences: An Integrated Approach,
   by J. Trefil and R. Hazen (Wiley, 2012).

11.1.2   Intermediate

1. The Computational Beauty of Nature, by G. Flake
   (MIT Press, 1998).

2. The Self-Made Tapestry, by P. Ball
   (Oxford University Press, 1999).

3. 'Exploring Complex Networks', by S. Strogatz,
   Nature 410 (2001) 268.

4. Growing Artificial Societies, by J. Epstein and R. Axtell
   (Brookings Institution Press, 1996).

5. The Web of Life, by F. Capra (Anchor Books, 1996).

6. NetLogo: Free agent-based model builder at
   https://ccl.northwestern.edu/netlogo/

11.1.3   Specialised

1. Nonlinear Dynamics and Chaos, by S. Strogatz
   (Westview Press, 2000).

2. Nonlinear Physics with Maple for Scientists and Engineers,
   by R. Enns and G. McGuire (Birkhauser, 2000).

3. Nonlinear Dynamics, Mathematical Biology, and Social Science,
   by J. Epstein (Addison Wesley, 1997).

4. Evolutionary Dynamics, exploring the equations of life,
   by M. Nowak (Harvard 2006).

5. Signs of Life, by R. Sole and B. Goodwin (Basic Books, 2002).

6. 'The structure and function of complex networks',
   by M. Newman, SIAM Review 45, 167 (2003).
   Available from http://arxiv.org/abs/cond-mat/0303516/

11.2   Using Complexity

How can you use the ideas from previous chapters, or the references above, to explore your area of interest?

For a start, try to re-look at what you do and know from a different perspective — that of complex systems. Are there hints of self-organisation? Could some processes be modelled simply?

Remember that within each discipline or problem there are hierarchies of complexity: It is rarely useful to study all levels at one go. For example, are you interested in the whole human body as a system, or is your interest mainly in the heart? Choose a model and language appropriate to that hierarchical level.

It is prudent to familiarise yourself, at least qualitatively, with some previous approaches to your particular problem. It is also helpful to draw inspiration from complex systems outside your discipline (recall how, for example, ant colonies inspired computer scientists).

Study simplified models to develop insight, before attempting something more involved.

Remember to have fun in the process.

11.3   Research in Complexity

If you are determined to study a particular complex system in depth, you would need to look at the research (technical) literature. The pointers below are for beginning researchers, who may also not have access to a university's library database, but do have access to the WWW.

Suppose, for example, you are interested in modelling vehicular traffic flow along a highway. You might start by typing some keywords in Google Scholar (scholar.google.com, the specialised, but still free, option on Google search): 'traffic flow model'.

If you are interested more in agent-based models, then add 'agent-based' as another keyword to delimit your search.

It is often useful to obtain an overview of the field by first reading a review of the subject , so add 'review' to narrow your search.

You might find some of the articles freely available on the web as pdf files or in other formats. Googling the author's name and the paper's title might lead you to the author's homepage, which might have other related material, sometimes even a copy of the paper you seek.

If the article is not freely available, but you think it might be useful to you (based on the title and abstract), you can try contacting the author directly through email to request a copy.

Once you have one or two interesting papers, attempt to first get the 'big picture' by reading the abstract, the introduction, and conclusion of the papers, rather than immediately digging into the details in the main body of the papers. Some papers might be more accessible than others.

If you think you have found some good papers, you might also want to check the citations to that paper (for example, through Google Scholar), to get an idea of what other researchers think of your favourite paper. Remember, even peer-reviewed published papers can contain errors and mistaken conclusions.

Suppose that you have spent some weeks narrowing down your reading material. The next stage would be an in-depth study. If there is much technical material in the paper, you might need to obtain some familiarity with the necessary tools.

You could also try contacting the authors of the paper for suggestions of supplementary reading or on clarifying some details of the paper (do not expect a tutorial though).

It will NOT be easy to do all this by yourself if you are a beginner — it is better to do it in a group, preferably supervised by an experienced researcher.

11.4   Some Open Problems

Here are some problems that interest this author at time of writing:

1. Origin of Life

   Living matter contains DNA that allows information to be transmitted from one generation to another, with mutations facilitating evolution.

Living matter also contains crucial metabolic systems to sustain life.

If living matter originated from non-living matter, as many scientists hypothesise, how did it happen? What were the simpler intermediate steps in the evolution of the complex metabolic and replication systems?

2. Consciousness

It is one of those things for which various partly satisfactory definitions exist, such as 'self-awareness'.

Is consciousness an emergent property of the biological brain? How would one prove or disprove that? This leads us to the next question.

3. Digital Life and Consciousness

Could man-made computational systems reach sufficient complexity to be indistinguishable (in essence) from life and consciousness as we know them now?

4. Viruses

Some viruses, such as HIV, are remarkably adaptable. What strategy should be used to manage them?

5. Solar cells

At time of writing, commercially viable solar cells have a relatively low efficiency. How can this be overcome?

6. Mind over Matter

There are various suggestions, such as from the placebo effect and studies involving mindfulness practices, that the mind may be used to affect changes in the body. How does this work and how far can one go with this?

7. DNA

We have learnt much, but probably much remains to be unravelled.

You will find many other open problems through the references on this book's website.

11.5   Exercises

1. Attempt to define consciousness precisely. Do non-human animals have consciousness? How can you tell?

2. Define 'life' (living matter), emphasising the essential characteristics. Would your definition encompass potential new forms that might be encountered in the future?

3. What ingredients would be needed for life, similar to what we know, to potentially exist elsewhere in the Universe? Are those conditions likely to be met?

4. Identify one acute problem facing your organisation or city. Gather data to quantify the problem. Analyse probable causes and potential solutions. Attempt to create a model that would simulate how your proposed solution would alleviate the problem. Present your proposal to a critical audience and improve your model, if necessary or possible, based on the feedback received. Obtain approval to test your proposal. Write up your results and submit them to a peer-reviewed journal for publication. Good luck!

Index

Notes

Notes

Notes

Simplicity in Complexity
Companion website: www.simplicitysg.net/books
Facebook: www.fb.com/simcomty

The Author

Dr. Rajesh R. Parwani

specialises in quantum theory and cosmology. He has mentored
research students from high-school to graduate school, and taught
university level modules in quantitative reasoning, multi-disciplinary
science and physics.
His other books are *Integrated Mathematics for Explorers* (with
Adeline Ng) and *Real World Mathematics* (with Dr. W.K. Ng).

Contact email: enquiry@simplicitysg.net

Print Edition SRI-2015-1A